Circular Economy for Sustainable Development

This book addresses the relationships among innovative technologies, methodologies, and processes and the different aspects of newly required digital and circular models that production systems and the economy must adopt and operate. It examines the latest innovations for the circularity of processes and end products, instilling the environmental consciousness of recycling and the ecosystem service valuation (ESV). This multidisciplinary approach, and the many case studies included in the book, helps readers understand different methodologies used when applying and implementing the principles of circular economy and green innovation at a practical level.

Features

- Provides the most recent developments and innovations for a circular economy.
- Discusses practical examples of how to understand innovation processes and frameworks that contribute to a circular economy.
- Identifies circular water management, circularity evaluation, innovation, and circularity standards.
- Includes practical case studies from different regions, showcasing successful management strategies and lessons learned.
- Highlights green entrepreneurship, strategy, and innovation management.

This book is for professionals, researchers, academics, and students in the fields of energy planning, sustainable development, environmental economics, business management, and entrepreneurship.

Circular Economy for Sustainable Development

Applications and Case Studies

Edited by
Konstantinos G. Aravossis,
Grigorios L. Kyriakopoulos, and Eleni Strantzali

CRC Press
Taylor & Francis Group
Boca Raton London New York

CRC Press is an imprint of the
Taylor & Francis Group, an **informa** business

Designed cover image: © iStock

First edition published 2026
by CRC Press
2385 NW Executive Center Drive, Suite 320, Boca Raton FL 33431

and by CRC Press
4 Park Square, Milton Park, Abingdon, Oxon, OX14 4RN

CRC Press is an imprint of Taylor & Francis Group, LLC

ISBN: 978-1-041-10407-0 (hbk)
ISBN: 978-1-041-10408-7 (pbk)
ISBN: 978-1-003-65488-9 (ebk)

DOI: 10.1201/9781003654889

Typeset in Palatino
by Apex CoVantage, LLC

Contents

Preface

Nowadays, the adoption of the circular economy is inseparably perceived as the evolving model of linear economy, in which the "take-make-waste" routes imposed the extraction, use, and then discard of resources in an unsustainable way. Subsequently, long-term depletion of natural resources and ongoing waste accumulation globally have excessively surpassed the environmental capacity of ecosystems. In response to this vicious cycle of environmental depletion, circular economy offers a regenerative alternative method that emphasizes to the continuous reduction, reuse, and recycling of products and materials, thus achieving a caring sustainable development for resource efficiency, pollution reduction, and climate change/crisis mitigation. This closed-loop system of circular economy further supports sustainable consumption and production, praising an otherwise depreciated matter for successive cycles of use instead of uncontrolled disposal of waste and wastewater effluents. Subsequently, coping with environmental and socio-economic challenges is simultaneously complying with the contemporary priority actions inspired by the United Nations Sustainable Development Goals (SDGs).

In this multi-parametric context, *Circular Economy for Sustainable Development: Applications and Case Studies* covers a broad variety of contemporary research, applications, and case studies in a wide geographic context. The novel attribute of this book is the combination of broad theoretical coverage, coupled with technologies, innovations, and case studies, thus achieving a meaningful framework that integrates "circular economy" from the (sometimes detached) university arena and familiarizing it with everyday and real-world environmental problem-solving toward achievement of "sustainable development."

The carefully selected chapters, written by highly environment-specialized academic experts and shifting from the linear to the circular model, offer insights into how governments, institutions, local authorities, field scientists, practitioners, higher education students, and university instructors can create more inclusive strategies for a sustainable environmental future. This is a timely, valuable book that pairs "circular economy" with "sustainable development" in a familiarized and insightful manner to the general readership, surveilling both entities towards a cleaner and supportive socio-economic level. It will be the editors' pleasure and happiness if readers find this book fascinating as it relates to their own research works, or just to be kept informed about contemporary emerging environmental issues, methods, innovations, and technologies.

The editors
Prof. Konstantinos G. Aravossis, Dr. Grigorios L. Kyriakopoulos,
Assistant Prof. Eleni Strantzali

About the Editors

Konstantinos G. Aravossis is a professor and UNESCO Chairholder on Green Innovation and Circular Economy at the National Technical University of Athens. Prof. Aravossis is a Vice President and Waste Management BoD Chairman Regulatory Authority for Waste, Energy and Water – RAAEY, and a former Secretary General for Forestry at the Greek Ministry of Environment and Energy. Prior to that, he was also Secretary General of Natural Environment & Water. He has been a professor at Athens University of Economics since 1997. He graduated in Mechanical Engineering from the Technical University of Aachen, Germany with a M.Sc. in management science from Imperial College in London and a Ph.D. in operations research from the National Technical University of Athens. He has organized 12 international conferences, has published more than 100 research studies in journals and books, and has coordinated numerous national and international research projects.

Grigorios L. Kyriakopoulos is a teaching and research associate at the School of Electrical and Computer Engineering, National Technical University of Athens, Greece. He holds two Ph.D. degrees, in low-carbon economy from DUTH, Greece and in chemical engineering from NTUA, Greece. He has authored and co-authored more than 140 papers, 35 book chapters, and 30 conference papers. He is the reviewer of more than 7,000 manuscripts at 525 journals. He is also an associate researcher at the UNESCO program on Green Innovation and Circular Economy at the National Technical University of Athens. His research interests are environmental systems and remediation, renewable energy sources, and circular economy, among others.

Eleni Strantzali is an assistant professor at the Department of Naval Architecture at University of West Attica, Greece, with a specialization in techno-economic analysis and supply chain in maritime transport. She graduated in mechanical engineering from NTUA, Greece and has a Ph.D. in industrial management and operational research from the same university. She is an associate researcher at the UNESCO program on Green Innovation and Circular Economy at the National Technical University of Athens.

Contributors

Konstantinos G. Aravossis
UNESCO Chair on Green Innovation and Circular Economy
National Technical University of Athens
Athens, Greece

Evangelos Baltas
Department of Water Resources and Environmental Engineering
National Technical University of Athens
Zografou, Greece

Epaminondas D. Banos
BOD member of Chios Marine Club
Piraeus, Greece

Pantelis Broukos
KOIOS Research and Innovation Center of Excellence
Department of Electrical and Computer Engineering
University of Cyprus
Greece
and
UNESCO Chair on Green Innovation and Circular Economy
School of Mechanical Engineering
National Technical University of Athens
Athens, Greece

Yassine Charabi
Department of Geography
Kuwait University
Kuwait City, Kuwait

Sotiria Dimitrellou
Department of Naval Architecture
School of Engineering
University of West Attica
Athens, Greece

Jan W. Dobrowolski
Committee on Geoinformatics and Environmental Engineering
Polish Academy of Sciences, Krakow Branch
Krakow, Poland

Isidoros Iakovidis
Department of Naval Architecture
School of Engineering
University of West Attica
Athens, Greece

Christos Kalantzis
UNESCO Chair on Green Innovation and Circular Economy
National Technical University of Athens
Athens, Greece

Konstantinos Kalkanis
Department of Electrical and Electronics Engineering
University of West Attica
Athens, Greece

Grigorios L. Kyriakopoulos
School of Electrical and Computer Engineering
National Technical University of Athens, Zografou Campus
Athens, Greece

Georgios A. Livanos
Department of Naval Architecture
School of Engineering
University of West Attica
Athens, Greece

Christos Nikoloudis
UNESCO Chair on Green Innovation and Circular Economy
National Technical University of Athens
Athens, Greece

Dimitrios-Nikolaos S. Pagonis
Department of Naval Architecture
School of Engineering
University of West Attica
Athens, Greece

Mathaios Panteli
KOIOS Research and Innovation Center of Excellence
Department of Electrical and Computer Engineering
University of Cyprus
Greece

Athanasios-Foivos Papathanasiou
Department of Water Resources and Environmental Engineering
National Technical University of Athens
Zografou, Greece

Ioannis Platanitis
Department of Water Resources and Environmental Engineering
National Technical University of Athens
Zografou, Greece

Constantinos S. Psomopoulos
Department of Electrical and Electronics Engineering
University of West Attica
Athens, Greece
and
Earth Engineering Center
Columbia University
New York, USA

Ioannis Sebos
School of Chemical Engineering
National Technical University of Athens, Zografou Campus
Athens, Greece

Eleni Strantzali
Department of Naval Architecture
School of Engineering
University of West Attica
Athens, Greece

Vasilios Viziryiannakis
Thalis Environmental Services
Athens, Greece
and
Motor Oil S.A.
Athens, Greece

Tadeusz Żaba
Krakow Water
Cracow University of Technology
Faculty of Environmental Engineering and Energy
Krakow, Poland

1

The Role of Industrial Symbiosis in the Circular Economy

Konstantinos Kalkanis, Vasilios Viziryiannakis, and Constantinos S. Psomopoulos

1.1 Introduction

Material flow management is made considerably more challenging by multi-layered bills of materials. The transition to circular economy (CE) material flows requires significant organisational changes, ranging from reorganising supply chain processes to fostering a sustainable mindset among product designers. Decisions made by operations managers will inevitably have an impact on the shift to CE models since managing reverse cycles and processes of reuse and remanufacturing require systemic changes across a wide range of domains, from product development to production and supply chain management. Scholarly interest in the ideas of CE and industrial symbiosis (IS) has increased as a result of the current trend towards sustainable economic models. By encouraging behaviours that optimise resource efficiency, minimise waste, and improve economic and environmental resilience, both frameworks seek to overcome the drawbacks of the conventional linear "take–make–dispose" model. While IS more precisely refers to cooperative exchanges of materials, energy, water, and by-products between firms, frequently located within a shared geographical boundary, CE is generally understood to be a systemic approach to closing material and energy loops across value chains (Chertow, 2000, 2007).

In scholarly and policy discourse, the connection between CE and IS is becoming more widely recognised. According to scholars, IS offers concrete means of putting CE principles into practice, especially at the meso and regional levels (Neves et al., 2020; Castellet-Viciano et al., 2022). Circular strategies are being implemented in practice through the creation of eco-industrial parks (EIPs), the growth of waste-to-resource programmes, and the use of industrial ecology concepts. CE and IS, however, have different conceptual origins despite their substantial overlap: CE comes from economic and policy

DOI: 10.1201/9781003654889-1

viewpoints that emphasise system-wide changes, whereas IS has developed within the field of industrial ecology with a more technical and operational focus (Fraccascia et al., 2017; Oughton et al., 2022).

Since the transition to circular business models necessitates adjustments in several areas, such as manufacturing, supply chain management, product development, and reverse cycle management, it influences the choices made by operations managers. Production scheduling issues are brought on by the CE's modifications, which include disassembly and the reverse flows required for remanufacturing or reuse (Acerbi & Taisch, 2020). Closed loop solutions can only be implemented with full supply chain management cooperation. Thus, supply chain management and logistics need to adjust to the new economic paradigm.

Partnerships between businesses are therefore crucial because they promote cooperation and the sharing of information. A small number of companies have already attained a high degree of integration, allowing them to create industrial eco-parks and symbioses that optimise both their direct and reverse flows. Significant improvements in reverse logistics have made it easier to recover trash from supply networks (Chertow, 2000, 2007; Fraccascia et al., 2017; Neves et al., 2020; Castellet-Viciano et al., 2022).

Managers creating a CE business model must determine the quantity of generated and virgin materials required, in addition to the anticipated quantity of recycled materials and parts for remanufacturing. Reverse logistics and the CE are closely related in terms of the technological cycle (material circularity and repair). Furthermore, both have something to do with the idea of sustainability. Figure 1.1 presents the concept of IS through a simplified scheme.

The broad concept of a creative, regenerative, and constructive economy is embodied by industrial sustainability. As a result, "industrial sustainability" refers to a collection of presumptions, guidelines, and analytical instruments for the productive system that seek to eradicate resource waste, make prudent use of resources and raw materials, match output to customer demands, and reverse logistics. IS, EIPs, and industrial ecology and its instruments are a few examples of transformative solutions that need to be used to address issues associated with CE growth and transition (Acerbi & Taisch, 2020; Chertow, 2000, 2007; Fraccascia et al., 2017).

The main practices of IS are cleaner production and ecodesign, which aim to improve process performance and efficiency while reducing waste production and environmental impact throughout a product's life cycle. The exchange of residues and byproducts in planned complexes of co-located manufacturing plants is the foundation of this strategy. One of the main components of reverse logistics is IS, which adds value to the initial inputs/resources and leads to mutually profitable transactions and more efficient processes because of the increased intensity of use of resources found in an industrial or urban complex (Acerbi & Taisch, 2020; Chertow, 2000, 2007).

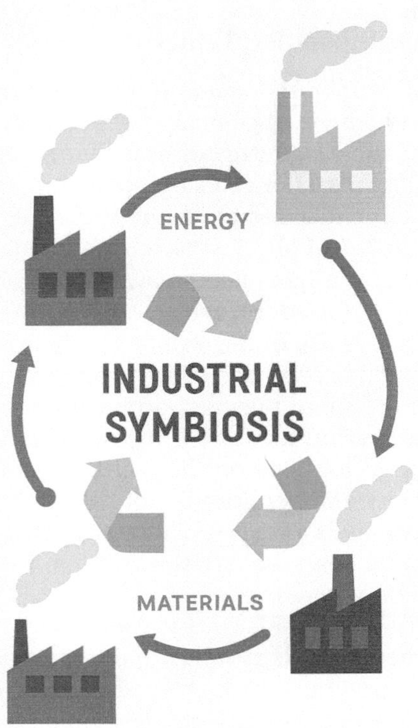

FIGURE 1.1
Basic presentation of the concept of industrial symbiosis.

Information and communication technology is the collective term for a collection of integrated technical resources that integrate hardware, software, and telecommunications to enable communication and automate business processes. The CE encourages the development of methods for gathering and processing municipal waste by fostering a reuse and recycling culture through the use of industry 4.0 and three-dimensional (3D) printing technologies (Lieder & Rashid, 2016).

Knowledge generation, knowledge sharing, and knowledge application are three ways that information systems offer various tools for teamwork. As a result, the digital platform facilitates data transmission through information systems that employ new technologies, procedures, and solutions for both direct and reverse logistics. Industrial ecology integrates the relationship between business and the environment by facilitating process design and providing tools such as material flow analysis, ecodesign, and cleaner production to help businesses implement sustainability.

1.2 Industrial Symbiosis Explained

The term "industrial symbiosis" refers to the ways in which a network of diverse organisations can enhance business and technical procedures, promote eco-innovation and long-term cultural change, and develop and share mutually beneficial transactions. Geographical proximity is frequently linked to IS, but it is neither required nor sufficient, and neither is a sole emphasis on the exchange of physical resources. To maximise the benefits of co-location, strategic planning is necessary. When IS is applied to business operations, resources are used, recovered, and redirected for reuse, which prolongs their productive use in the economy (Chertow, 2000, 2007; Fraccascia et al., 2020; Neves et al., 2020; Castellet-Viciano et al., 2022).

As a result, there are more business opportunities, less demand on the planet's resources, and a foundation for the development of a CE. A subset of industrial ecology, IS focusses on the exchange of materials and energy. Although the simplicity and applicability of this paradigm have been questioned, industrial ecology is a relatively new field that is based on a natural paradigm and asserts that an industrial ecosystem may behave similarly to the natural ecosystem, where everything is recycled (Chertow, 2000, 2007; Fraccascia et al., 2020; Neves et al., 2020; Castellet-Viciano et al., 2022).

One way that industrial ecology helps to integrate environmental preservation and economic growth is through eco-industrial development.

Some of the examples of eco-industrial development are the following.

- Circularity of single resource (single material and/or energy exchange).
- Greenfield eco-industrial development (geographically confined space).
- Brownfield eco-industrial development (geographically confined space).
- Eco-industrial network (no strict requirement of geographical proximity).
- Virtual eco-industrial network (networks spread in large areas, e.g., a regional network)
- Networked eco-industrial system (macro-level developments with links across regions).

Through the physical exchange of materials, energy, water, and/or byproducts, IS involves historically distinct industries working together to gain a competitive edge. Cooperation and the synergistic opportunities provided by geographic proximity are the keys to IS. Notably, the National Industrial Symbiosis Programme's research and published activities in the UK have investigated and empirically tested this definition, as well as the stated key aspects of IS, namely the role of collaboration and geographic proximity, in its variety of forms (Chertow, 2000, 2007).

Systems of IS optimise energy and material use at levels higher than any one process could achieve on its own. Over a long period of time, IS systems like

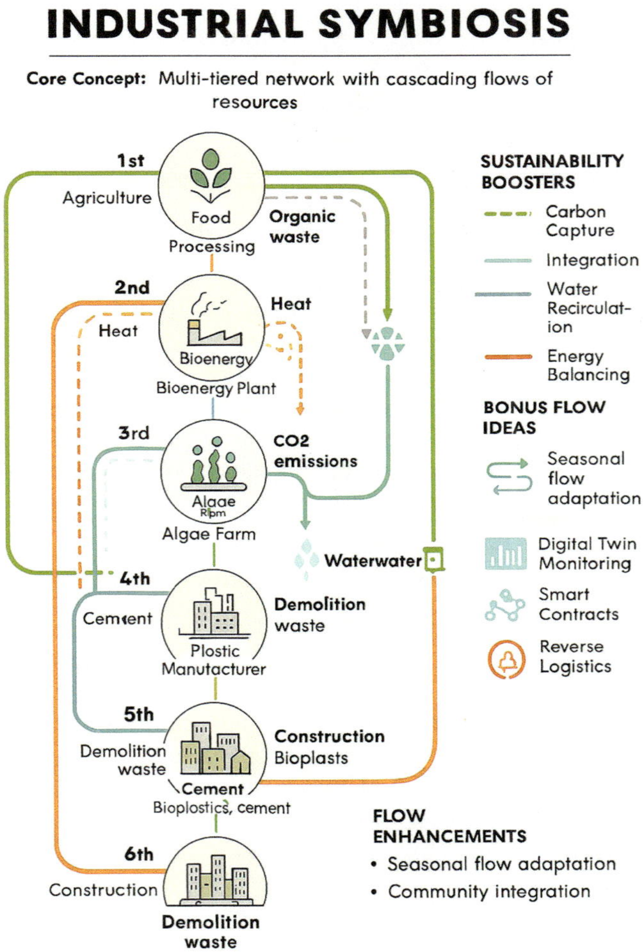

FIGURE 1.2
A theoretical concept of industrial symbiosis created by the authors, presenting the possibilities of arising from its adoption in industrial processes.

the web of materials and energy exchanges among businesses in Kalundborg, Denmark, have developed naturally from a number of micro-innovations. However, from the standpoint of a macro planner, the engineered design and implementation of such systems on a relatively short time scale proves difficult. Addressing this, Figure 1.2, created by the authors, provides the potential materials, energy, and water flows between completely different industries, including communities, also providing potential approaches for optimisation.

Information about available by-products is frequently hard to come by. Usually not traded or listed on any kind of exchange, these by-products are regarded as waste. This type of waste trading is handled by a limited number of specialised waste marketplaces.

1.3 Defining Circular Economy Through Industrial Symbiosis

A collection of overlapping theoretical underpinnings underpins the relationship between CE and IS, guiding both conceptual development and real-world application. In order to shift from the conventional linear economic model of extraction, production, consumption, and disposal to a regenerative, restorative, and resource-efficient paradigm, CE and IS both aim to rethink production and consumption systems. The theoretical foundations of CE and IS are examined in this section, along with their intersections and unique viewpoints, and how these frameworks influence one another.

1.3.1 Theoretical Roots of the Circular Economy

Several schools of thought have influenced the development of the CE concept. These include natural capitalism, biomimicry, performance economy, cradle-to-cradle design, regenerative design, and industrial ecology (Geissdoerfer et al., 2017). In order to separate economic growth from environmental degradation, CE places a strong emphasis on the ongoing cycling of resources through restorative and regenerative processes (Korhonen et al., 2018). It suggests that in order to minimise waste, preserve natural capital, and foster economic innovation, materials should be kept in closed-loop systems for as long as feasible.

The concept of closing, slowing, and narrowing resource loops has a significant influence on CE theory (Bocken et al., 2016). While slowing loops focus on prolonging product lifespans, narrowing loops seeks to minimise resource input through efficiency, and closing loops recycles materials back into the economy. CE is especially compatible with IS because of its systemic perspective, which operationalises these concepts at the industrial level through material and energy exchanges.

1.3.2 Theoretical Roots of Industrial Symbiosis

IS has its roots in the larger field of industrial ecology, which studies industrial systems using systems thinking and ecological metaphors. The Kalundborg industrial ecosystem in Denmark – where businesses trade water, energy, and byproducts to replicate the nutrient cycles of natural ecosystems – popularised the idea of IS (Chertow, 2000).

Theoretically, IS is based on ecological analogies and systems thinking. According to Doménech and Davies (2011), it presents industries as interdependent parts of an ecosystem in which the waste and byproducts of one become useful inputs for another. Instead of focussing on individual entities, IS promotes the development of synergy and the optimisation of resource use across networks of firms. By encouraging diversity, redundancy, and resource efficiency, it helps industrial networks become more resilient and adaptable (Fraccascia et al., 2021).

1.3.3 Convergences Between CE and IS

Despite having similar goals and theoretical underpinnings, CE and IS function at different scales and with distinct emphasis. While IS concentrates on meso-level cooperation between businesses and industries, CE offers the broad policy and philosophical framework, stressing macro-level changes (Chertow, 2000; Fraccascia et al., 2020).

Several points of convergence are evident, such as the following.

- **Systems orientation:** Both CE and IS adopt systemic perspectives, recognising the interconnectedness of industrial and natural systems.
- **Resource valorisation:** Both frameworks aim to maximise the utility of resources through reuse, recycling, and repurposing.
- **Innovation and competitiveness:** CE and IS promote innovation in business models, technologies, and collaborative practices as pathways to sustainability.

1.3.4 Distinctions Between CE and IS

CE and IS are different, even though they overlap. CE is more comprehensive, covering business models at all scales, consumer behaviour, policy, and product design (Korhonen et al., 2018). Government initiatives and international sustainability agendas are frequently used to promote it. IS, on the other hand, has a more limited scope and focusses exclusively on resource-sharing partnerships between firms. It is practice-oriented and based on localised, context-specific interactions.

These differences demonstrate how the two frameworks work in tandem: CE establishes the goal, while IS offers the practical tools to reach it.

1.3.5 Mutual Reinforcement

It is becoming more widely acknowledged that attaining sustainable industrial systems requires the integration of CE and IS. Through concrete instances of resource cycling, teamwork, and creativity, IS illustrates the viability of CE principles in real-world settings. By integrating IS into broader business and policy frameworks, CE simultaneously broadens the applicability of IS beyond EIPs (Korhonen et al., 2018).

As a result, the theoretical underpinnings of IS and CE are not only compatible but also reinforce one another. While IS offers the operational and practical implementation pathways, CE provides the normative and policy context.

1.4 Methodological Approaches in the Literature

Over the past 20 years, there has been a significant increase in the study of IS and CE, using a variety of methodological approaches to comprehend, assess, and direct their development. A characteristic of CE–IS research that reflects the interdisciplinary nature of the field is methodological diversity. The main research approaches are examined in this section, along with their advantages and disadvantages. These approaches include life cycle assessments (LCAs), modelling techniques, case study analyses, systematic reviews, and hybrid frameworks (Doménech & Davies, 2011).

1.4.1 Systematic Reviews and Bibliometric Analyses

A common method for synthesising the body of knowledge on CE and IS is systematic literature reviews. In their thorough analysis of the IS literature, Neves et al. (2020) noted important themes, patterns, and areas in need of further study. Similar to this, Fabris et al. (2022) mapped the evolution of knowledge using bibliometric techniques, emphasising the growing overlap between IS and CE scholarship.

Bibliometric analyses provide important information about how ideas change over time, which areas are the focus of most research, and which areas are still unexplored. They demonstrate how, especially in Europe and East Asia, CE-IS research has expanded from a specialised area of scholarly interest to a mainstream sustainability conversation.

1.4.2 Case Study Methodologies

The foundation of IS research continues to be case studies. The most famous example, frequently regarded as the founding example of IS, is the Danish industrial ecosystem of Kalundborg (Chertow, 2000). Case studies have examined a variety of contexts outside of Kalundborg, such as forestry-based networks (Simioni et al., 2024), agrifood clusters (Ruini et al., 2025) and construction waste valorisation (Yu et al., 2021).

These examples show how IS can take many different forms, ranging from digital platform-mediated networks and by-product valorisation to energy exchanges and water reuse. They also make clear how institutional, cultural, and policy contexts influence the dynamics of IS.

1.4.3 Network Analysis and Modelling Approaches

The use of network analysis to study IS dynamics and structures has grown in popularity. Researchers can examine network density, centrality, modularity, and resilience by considering firms as nodes and exchanges as links (Fraccascia et al., 2021). These metrics highlight important businesses or exchanges and provide insight into how resilient symbiotic systems are.

Additional insights are offered by modelling techniques such as agent-based modelling and input–output analysis. While agent-based models mimic firm behaviour and interactions within IS networks, input–output models capture material and energy flows at macro and meso scales. These methods highlight the strengths and weaknesses of IS systems in various contexts.

1.4.4 Life Cycle Assessment (LCA)

LCA is frequently used to measure how CE and IS initiatives affect the environment. Yu et al. (2021) illustrated the advantages of recycled concrete aggregates for the environment using LCA. Ruini et al. (2025) evaluated resource and greenhouse gas (GHG) emission reductions by applying LCA to agrifood IS systems.

LCA offers a strict framework for weighing trade-offs, identifying situations in which symbiosis might change environmental burdens rather than lessen them. Complementary methods are necessary because LCA frequently fails to capture systemic and social impacts.

1.4.5 Hybrid Approaches

To produce more thorough assessments, hybrid approaches integrate network analysis, input–output modelling, and LCA. Fraccascia et al. (2017), for example, highlighted the contributions of IS to systemic robustness by combining resilience analysis with environmental assessments. The growing understanding that no one approach can adequately represent the complexity of CE-IS systems is reflected in these hybrid frameworks. Rather, multi-method approaches are required to assess social, economic, and environmental aspects at the same time.

1.4.6 Governance and Policy Analysis

Methodologies focused on governance look at how policies either support or impede the development of IS. While Oughton et al. (2022) concentrated on governance capacity and flexibility in IS systems, Castellet-Viciano et al. (2022) employed institutional analysis to investigate EIPs. These methods emphasise how important institutional, regulatory, and policy frameworks are in determining the viability and success of IS projects. By placing IS in larger political and social contexts, they enhance technical evaluations.

1.4.7 Digital and Data-Driven Tools

Digital tools are incorporated into methodological approaches in recent scholarship. Digital platforms provide fresh perspectives on IS dynamics by acting as both research objects and data sources (Sgambaro et al., 2024). Large datasets on transactions and exchanges are produced by these platforms, allowing predictive analytics and machine learning to find possible synergies.

The ability to map, monitor, and optimise symbiotic networks is thus improved by data-driven tools, which constitute a frontier in CE-IS research.

1.4.8 Strengths and Limitations

Each methodological approach offers strengths and faces limitations, as follows.

- Reviews consolidate knowledge but cannot capture emerging practices in real time.
- Case studies provide rich insights but are often context-specific and difficult to generalise.
- Network models highlight structures but may oversimplify sociocultural dynamics.
- LCA quantifies environmental impacts but struggles with systemic complexity.
- Hybrid approaches offer comprehensiveness but require significant data and expertise.
- Governance analyses contextualise IS but may underemphasise technical aspects.
- Digital tools provide new opportunities but raise concerns about data access and trust.

1.4.9 Toward Methodological Integration

By fusing traditional and digital tools, technical and social viewpoints, qualitative and quantitative approaches, the field is heading towards methodological integration. The intricacy of CE-IS systems and the demand for comprehensive comprehension are reflected in this integrative trend (Fraccascia et al., 2020). In order to fill in the gaps in scope, scale, and inclusivity, future research will probably keep combining different approaches.

1.4.10 Discussion

1.4.10.1 Industrial Symbiosis as a Driver for Circular Economy Implementation

As a useful tool for closing, slowing, and narrowing resource loops, IS is essential to operationalising the ideas of CE. The broad concept of systemic resource efficiency and waste reduction is provided by CE, while the concrete procedures, partnerships, and interactions that make these principles a reality at the regional and industrial levels are provided by IS.

1.4.10.2 Waste Valorisation and By-Product Exchange

Waste valorisation is one of IS's fundamental contributions to CE. One company's waste streams and by-products are converted into useful resources for another, preventing landfilling and lowering the need for virgin inputs. According to studies, material flows like ash, gypsum, and food waste are valued in industries like forestry (Simioni et al., 2024), agrifood (Ruini et al., 2025), and construction (Yu et al., 2021).

1.4.10.3 Energy Symbiosis and Utility Sharing

IS promotes energy symbiosis, which includes heat cascading, steam sharing, and renewable integration, in addition to material exchanges. The classic example is still Kalundborg's steam network, but more recent EIPs in China, South Korea, and Europe show how energy exchanges promote CE by boosting productivity and reducing emissions (Castellet-Viciano et al., 2022; Neves et al., 2020).

1.4.10.4 Eco-Industrial Parks as CE Hubs

The IS–CE nexus, in which groups of industries work together to minimise waste and maximise resources, is embodied by EIPs. By incorporating circularity principles into governance, resource management, and infrastructure, Castellet-Viciano et al. (2022) demonstrate how EIPs directly correspond with CE strategies. These parks – which frequently gain from encouraging laws and regulations – are prime examples of CE in action.

1.4.10.5 Business Model Innovation

Product-as-a-service, resource recovery businesses, and symbiosis-based service platforms are some of the new business models that IS makes possible and are essential to CE. These models are being accelerated by digital tools, with platforms that make it easier to match waste suppliers and users (Sgambaro et al., 2024).

1.4.10.6 Policy and Institutional Integration

The use of IS as a tool to carry out CE policies is becoming more widely acknowledged by governments. Using financial, regulatory, and informational tools to promote cooperation, policy frameworks in China and the EU specifically incorporate IS into circular strategies (Oughton et al., 2022).

1.4.10.7 Socio-Environmental Co-Benefits

IS produces broader socio-environmental benefits that are in line with CE objectives, such as decreased GHG emissions, improved community resilience, resource efficiency, and job creation. Research highlights that IS supports both regional development and industrial sustainability (Neves et al., 2020).

1.4.10.8 Limitations and Challenges

Although IS facilitates CE, scaling is still difficult. Information asymmetry, high transaction costs, problems with trust, and incompatibilities between technologies are some of the obstacles. Digitalisation, robust governance, and integrative policy design are necessary to overcome these.

1.5 Barriers and Challenges

Despite being widely acknowledged as a crucial facilitator of CE strategies, IS encounters numerous obstacles in its implementation. Institutional, economic, technical, organisational, informational, and social aspects are all included in these multifaceted challenges. Designing interventions that facilitate the shift towards CE requires an understanding of these obstacles.

1.5.1 Institutional and Regulatory Barriers

The requirements of IS networks are frequently not met by institutional and regulatory frameworks. By-products may be categorised as hazardous by waste regulations, resulting in bureaucratic obstacles that deter trade (Oughton et al., 2022). Cross-regional cooperation is also hampered by a lack of uniformity among jurisdictions – especially in the EU, where member states implement CE principles in disparate ways.

1.5.2 Economic and Financial Barriers

Economic uncertainty presents significant challenges. The willingness of firms to participate in IS is limited by high transaction costs, uncertain return on investment, and market volatility. Smaller businesses frequently lack the funding necessary to upgrade their technology or invest in the infrastructure required for exchanges (Fraccascia et al., 2017).

1.5.3 Technical and Infrastructural Barriers

Exchanges may be impeded by technical incompatibilities between production processes and waste streams. Feasibility is further limited by inadequate infrastructure, including storage, transportation, and pipelines. Furthermore, before certain by-products can be used again, they might need to undergo expensive pre-treatment (Yu et al., 2021).

1.5.4 Organisational and Managerial Barriers

High levels of coordination between various actors are necessary for IS. Collaboration is frequently hampered by differences in corporate culture, risk perception, and strategic priorities (Doménech & Davies, 2011). It is challenging to maintain networks over time in the absence of committed managers or facilitators.

1.5.5 Information and Knowledge Barriers

One persistent problem is a lack of openness and information exchange. Businesses may be hesitant to reveal waste stream data because of mistrust or competitive concerns. The problem is made worse by a lack of knowledge about possible symbiotic opportunities. Despite adoption and trust barriers, digital platforms are becoming partial solutions (Sgambaro et al., 2024).

1.5.6 Social and Cultural Barriers

IS viability is influenced by cultural perspectives on waste and social acceptance. Reusing waste is stigmatised in some areas, while inter-firm cooperation is discouraged by cultural norms in others. It is crucial to establish trust and promote a cooperative culture (Neves et al., 2020).

1.5.7 Interconnectedness of Barriers

These barriers rarely happen alone; they often reinforce and interact with one another. Technical incompatibilities, for example, can raise expenses, which are made worse by regulatory limitations. In a similar vein, a lack of trust weakens network resilience by increasing informational barriers.

1.5.8 Strategies to Overcome Barriers

- **Policy reforms:** Clarifying legal definitions of by-products and harmonising waste regulations.

- **Economic incentives:** Subsidies, tax breaks, and funding schemes to reduce financial risks.
- **Technological innovation:** Pre-treatment technologies and modular infrastructure.
- **Facilitation and governance:** Appointing neutral coordinators or symbiosis facilitators.
- **Digital tools:** Platforms for transparent exchange of information.
- **Cultural shifts:** Awareness campaigns and community engagement to improve acceptance.

1.6 Benefits of IS Adoption

1.6.1 Environmental Benefits

Perhaps the most well-established and closely related to CE objectives are the environmental advantages of IS adoption. IS decreases raw material extraction, landfill waste, and GHG emissions by promoting resource sharing and waste valorisation (Yu et al., 2021; Simioni et al., 2024). Energy and water consumption can be significantly reduced in EIPs, where several businesses participate in synergistic exchanges (Neves et al., 2020).

Despite trade-offs, IS project LCAs frequently show net positive results. For example, Ruini et al. (2025) demonstrated that in agri-food IS networks, the environmental advantages of less waste exceeded the emissions linked to material exchanges and transportation. In a similar vein, Sgambaro et al. (2024) showed that Italian IS cases led to smaller environmental footprints in a variety of sectors.

1.6.2 Economic Benefits

Adoption of IS has a significant positive economic impact on participating businesses and areas. According to Fraccascia et al. (2017), cost savings are achieved through decreased disposal fees, lower material procurement costs, and shared infrastructure costs. The sale of waste materials as secondary resources turns liabilities into assets and creates new revenue streams.

IS supports resilience and economic diversification at the regional level. EIPs create competitive advantages by integrating circular strategies into local economies, according to Castellet-Viciano et al. (2022). In order to strengthen regional economic self-sufficiency, Simioni et al. (2024) emphasised the development of local bioenergy markets in forestry clusters.

1.6.3 Social Benefits and Employment

Adoption of IS has social benefits, as well. IS improves trust, collaboration, and social capital in regions by encouraging firm cooperation (Neves et al., 2020). Cleaner surroundings, better air and water quality, and less exposure to hazardous waste are all advantageous to communities (Ruini et al., 2025).

Another important social outcome is the creation of jobs. From digital platform management to waste valorisation technologies, IS projects frequently call for new competencies. These programmes improve workforce adaptability to CE-oriented economies and generate green jobs (Simioni et al., 2024).

1.6.4 Innovation and Competitiveness

By encouraging businesses to investigate novel applications for by-products and collaborate on the development of technologies for material recovery, IS networks promote innovation. This cooperative innovation makes regions leaders in CE transitions and increases firm-level competitiveness (Fraccascia et al., 2021).

Business models are not exempt from innovation. By-products of IS collaboration include digital marketplaces, product-as-a-service strategies, and shared service platforms (Oughton et al., 2022). These developments complement CE's focus on resource efficiency and systemic change.

1.6.5 Policy Benefits

From a policy standpoint, IS offers a concrete and quantifiable means of putting CE strategies into practice. Documented interactions and quantifiable decreases in emissions and resource consumption can serve as indicators of progress for policymakers (Castellet-Viciano et al., 2022). Effective IS initiatives serve as models, influencing and validating more general CE policy frameworks.

According to Ruini et al. (2025), by offering tangible instances of circular practices, IS initiatives can assist policymakers in overcoming the difficulty of operationalising CE at the local level. Consequently, this increases the pace of more extensive systemic changes.

1.6.6 Systemic and Resilience Benefits

Finally, one of CE's main objectives is systemic resilience, which IS helps to achieve. By diversifying resource inputs and outputs, symbiotic networks improve industrial systems' resilience to shocks (Fraccascia et al., 2017). IS offers substitute resource flows that preserve production stability in the event of disruptions like resource shortages or supply chain failures.

According to Neves et al. (2020), IS strengthens regional resilience by integrating businesses into collaborative networks. By promoting long-term stability and adaptability, these networks bring industrial systems into line with CE's goal of regenerative, sustainable economies.

1.7 Successful Case Studies

1.7.1 Kalundborg, Denmark

The Kalundborg Symbiosis, which is based on public–private partnerships and involves closed-loop energy, water, and material exchanges, is located in Kalundborg, Denmark. With a circular approach to production, Kalundborg Symbiosis aims to become the most successful IS in the world. A local collaboration in which partners supply, exchange, and repurpose resources to generate mutual value is known as an IS (Doménech & Davies, 2011; Shedroff, 2009).

In order to demonstrate some essential components of a CE at the local level, IS aims to establish loops of technical or biological materials while reducing leakage and waste in the loops. Beginning with the local water supply, this kind of symbiosis was established in Kalundborg in 1961. Because of the participants' belief that cooperation is simply smart business, it wasn't formally known as "industrial symbiosis" until 1989. Over the past 60 years, Kalundborg Symbiosis has developed naturally, with collaborative projects being developed when the business case and teamwork were advantageous and ending when the business case was not (Doménech & Davies, 2011; Shedroff, 2009).

Establishing a functional IS involves more than just cooperating for the benefit of one another and the environment; it also entails a constant pursuit of more efficient business practices. In light of this, Kalundborg has demonstrated that IS can function effectively as a hub for creative test and demonstration projects and offer excellent prospects for startups. Over the years, a number of projects have been finished, ranging from facilities for producing algae to the production of bio-ethanol, some of which are currently being carried out on a large scale. A number of startups have also set up shop in Kalundborg in order to take advantage of the city's innovative spirit and collaborative atmosphere (Doménech & Davies, 2011; Shedroff, 2009).

The foundational principles of Kalundborg Symbiosis – trust, confidentiality, openness, equality, and cooperation – have enabled all of the fruitful collaborations. In addition to fostering the symbiotic mindset among others on all geographic scales, from local to global, these values enable Kalundborg to revitalise and fortify its partnerships by tying together the flows of materials, water, and energy. According to the network diagram in Figure 1.3,

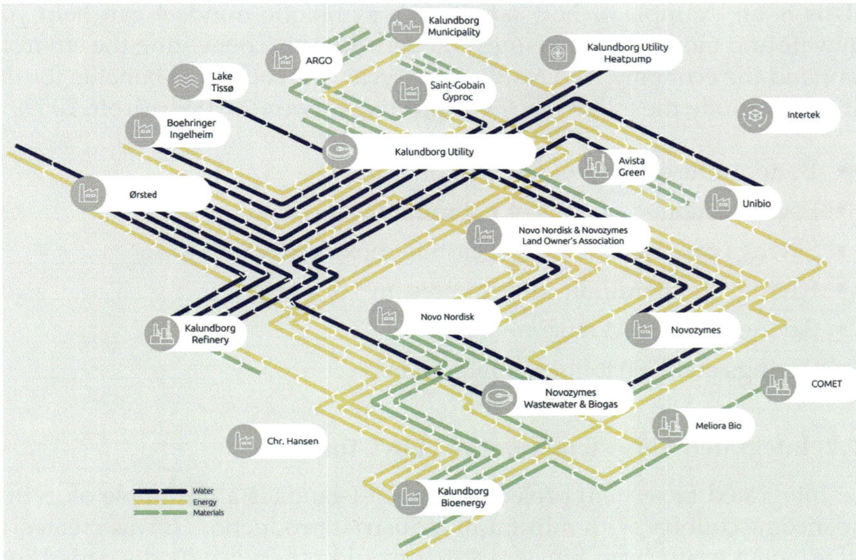

FIGURE 1.3
Kalundborg Symbiosis diagram.

Source: Kalundborg (2023)

Kalundborg Symbiosis is composed of 25 distinct streams, including material, energy, and water flows (Doménech & Davies, 2011; Kalundborg, 2023; Shedroff, 2009).

Three public sector organisations and six industrial organisations are the sources of the resource flows.

Nowadays, working together is second nature and every new choice made by each business takes into account the potential effects on the other partners, as well as Kalundborg overall. Shared values and the fact that none of the businesses are directly competing with one another support this way of thinking.

The transfer of high-temperature steam from Ørsted's combined heat and power plant, Asnaes Power Station, to numerous other symbiotic partners is another essential component and the most obvious and identifiable indication of the symbiotic collaboration. The characteristic green pipes that the steam exchange takes place through are the source of the visibility. The pipes that run through all of the industrial areas are what inspired Kalundborg's logo (Doménech & Davies, 2011; Shedroff, 2009).

Furthermore, this mutually beneficial project transformed Ørsted's business strategy from using excess steam as a by-product to generate electricity to using steam as the main product and the main source of revenue.

This is an example of how adopting a symbiotic mindset can help you innovate and modify your business model while also benefiting the environment and the economy. The result: every year, the combined benefits for the partners include the following (Doménech & Davies, 2011; Shedroff, 2009).

- Bottom-line savings of €24 million
- Socio-economic savings of €14 million
- Savings of 635,000 tonnes of carbon dioxide (CO_2)
- Savings of 3.6 million cubic metres (m^3) of water
- Savings of 100 GWh of energy
- Savings of 87,000 tonnes of materials

1.7.2 Integrated Biosystem Montfort Boys Town

The Integrated Biosystem project in the Fiji Islands is an example of IS in a low-income country with minimum industrial production. It was created to gather spent grain from breweries that would otherwise be discarded into the ocean, thereby suffocating coral reefs. The system turns brewery waste into high-value pig feed by using it as a substrate for the growth of mushrooms. After passing through an anaerobic digester, the pigs' waste is piped into fish ponds. By producing food for fish in all four trophic levels, this treatment enriches the water with nutrients (Chertow, 2000). Figure 1.4 presents the different streams in this IS case study.

Additionally, the waste produces fertile soil for vegetable cultivation. This project's unique feature is its integration of industrial and agricultural processes, which is reflected in its very name, "integrated biosystem" (Chertow, 2000).

1.8 Conclusions

This comprehensive review has compiled a wide range of research on the relationship between industrial symbiosis (IS) and the circular economy (CE), emphasising their methodological approaches, benefits, challenges, and emerging directions, as well as their conceptual alignments and practical synergies. When combined, the results show that IS is a useful means of achieving CE goals at all levels and in all sectors.

Several key insights emerge from this analysis, including the following.

- **Conceptual integration:** CE and IS share common goals of resource efficiency, waste minimisation, and systemic transformation. While CE provides the broad policy and philosophical framework, IS operationalises it through concrete exchanges and collaborations.

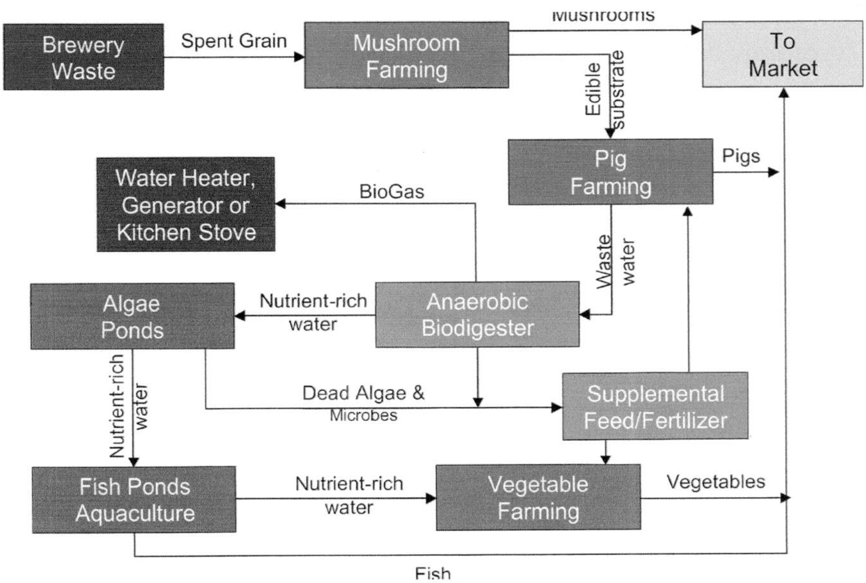

FIGURE 1.4
Flow diagram of the Montfort Boys Town Integrated BioSystem.

Source: Regenerated by the authors based on the original published in Chertow, 2000, Figure 3, p. 324

- **Methodological diversity:** Research employs a variety of methods – from case studies and network analysis to LCA and hybrid models – underscoring the interdisciplinary nature of CE–IS scholarship. Digitalisation is expanding methodological possibilities further.
- **Practical drivers:** IS acts as a catalyst for CE by enabling waste valorisation, energy efficiency, and circular business models. Its applications extend beyond traditional industries to include construction, agri-food, forestry, and urban systems.
- **Challenges:** Despite progress, IS integration faces barriers related to governance, economics, culture, data, and technical constraints. Overcoming these requires coordinated policy support, trust-building mechanisms, and technological enablers.
- **Benefits:** IS contributes not only to environmental performance but also to economic competitiveness, social well-being, and systemic resilience. These co-benefits strengthen the case for mainstreaming IS in CE transitions.
- **Assessment and indicators:** Effective monitoring remains a challenge. While LCA and network analysis provide useful insights,

there is a need for integrated, multidimensional frameworks that capture environmental, economic, social, and governance outcomes.

- **Emerging trends:** The field is evolving toward digitalisation, sectoral diversification, urban symbiosis, hybrid business models, and integration of social equity. These trends reflect both the maturity of the field and its readiness to scale globally.

Despite significant advances, the following notable gaps remain.

- Limited attention to **social and equity dimensions**, particularly in developing contexts.
- Insufficient integration of **digital technologies** into mainstream CE–IS frameworks.
- Lack of **comparative, longitudinal studies** assessing IS impacts across regions and sectors.
- Need for **global perspectives**, as most studies remain clustered in Europe and East Asia.
- Fragmentation of **indicators and metrics**, which hampers cross-case comparability.

Addressing these gaps will require collaboration among academics, policymakers, and industry stakeholders, as well as greater inclusivity in research agendas.

IS provides a useful tool for policymakers to convert CE goals into quantifiable results. Incentives for inter-firm cooperation, supportive infrastructures, and flexible regulations are necessary to support IS. IS offers industry chances for cost reduction, innovation, and resilience; however, achieving these advantages necessitates investing in trust-building mechanisms and a shift in culture. The challenge for researchers is to create solid frameworks and increase empirical data, especially in areas and industries that are under-represented.

The integration of CE and IS is a systemic change that reinterprets the relationships between ecosystems, industries, and societies rather than just being a technical exercise. By converting abstract circularity principles into concrete, quantifiable, and scalable practices, IS exemplifies the collaborative spirit of CE. IS will become more and more important in forming resilient, inclusive, and regenerative industrial ecosystems as the world's sustainability problems worsen.

In summary, the literature makes a compelling case for the necessity of IS in order to operationalise the CE. In order to improve efficiency and transparency, future research and practice should concentrate on creating inclusive and flexible indicators, encouraging multi-scale collaborations, and utilising digital technologies. By doing this, IS and CE can work together to create a more equitable and sustainable industrial future.

References

Acerbi, F., & Taisch, M. (2020). A literature review on circular economy adoption in the manufacturing sector. *Journal of Cleaner Production*, 273, 123086. https://doi.org/10.1016/j.jclepro.2020.123086

Bocken, N. M. P., de Pauw, I., Bakker, C., & van der Grinten, B. (2016). Product design and business model strategies for a circular economy. *Journal of Industrial and Production Engineering*, 33(5), 308–320. https://doi.org/10.1080/21681015.2016.1172124

Castellet-Viciano, L., Hernandez-Chover, V., Bellver-Domingo, Á., & Hernández-Sancho, F. (2022). Industrial symbiosis: A mechanism to guarantee the implementation of circular economy practices. *Sustainability*, 14, 15872. https://doi.org/10.3390/su142315872

Chertow, M. R. (2000). Industrial symbiosis: Literature and taxonomy. *Annual Review of Energy and the Environment*, 25, 313–337. https://doi.org/10.1146/annurev.energy.25.1.313

Chertow, M. R. (2007). "Uncovering" industrial symbiosis. *Journal of Industrial Ecology*, 11, 11–30. https://doi.org/10.1162/jiec.2007.1110

Doménech, T., & Davies, M. (2011). Structure and morphology of industrial symbiosis networks: The case of Kalundborg. *Ecological Economics*, 70(2), 603–611. https://doi.org/10.1016/j.ecolecon.2010.11.010

Fabris, M. M., Sabbadini, F. S., & Costa, K. A. (2022). A quantitative indicator for assessing the degree of maturity of industrial symbiosis. Research, Society and Development, 11(6), e43911629348. https://doi.org/10.33448/rsd-v11i6.29348

Fraccascia, L., Giannoccaro, I., & Albino, V. (2017). Efficacy of landfill tax and subsidy policies for the emergence of industrial symbiosis networks: An agent-based simulation study. *Waste Management*, 59, 262–273. https://doi.org/10.1016/j.wasman.2016.10.001

Fraccascia, L., Giannoccaro, I., & Albino, V. (2020). Industrial symbiosis and circular economy: The role of the firm. *Journal of Cleaner Production*, 268, 122321. https://doi.org/10.1016/j.jclepro.2020.122321

Fraccascia, L., Sabato, A., & Yazan, Δ. M. (2021). An industrial symbiosis simulation game: Evidence from the circular sustainable business development class. *Journal of Industrial Ecology*, 25(6), 1688–1706.

Geissdoerfer, M., Savaget, P., Bocken, N. M. P., & Hultink, E. J. (2017). The circular economy – A new sustainability paradigm? *Journal of Cleaner Production*, 143, 757–768. https://doi.org/10.1016/j.jclepro.2016.12.048

Kalundborg. (2023). Symbiosis infographic. Available in https://www.symbiosis.dk/wp-content/uploads/2023/05/Streams_WEB_EN_WONumbers_05-1.gif

Korhonen, J., Honkasalo, A., & Seppälä, J. (2018). Circular economy: The concept and its limitations. *Ecological Economics*, 143, 37–46. https://doi.org/10.1016/j.ecolecon.2017.06.041

Lieder, M., & Rashid, A. (2016). Towards circular economy implementation: A comprehensive review in context of manufacturing industry. *Journal of Cleaner Production*, 115, 36–51. https://doi.org/10.1016/j.jclepro.2015.12.042

Neves, A., Godina, R., Azevedo, S. G., & Matias, J. C. O. (2020). A comprehensive review of industrial symbiosis. *Journal of Cleaner Production*, 247, 119113. https://doi.org/10.1016/j.jclepro.2019.119113.

Oughton, C., Kurup, B., Anda, M. et al. (2022). Industrial symbiosis to circular economy: What does the literature reveal for a successful complex industrial area? *Circular Economy and Sustainability*, 2, 1317–1344. https://doi.org/10.1007/s43615-022-00153-1

Ruini, A., Bundgaard, A. M., Mosgaard, M. A., Løkke, S., & Pizzol, M. (2025). Normative and empirical solutions to the upcycling dilemma. *Sustainable Production and Consumption*, 54, 466–475. https://doi.org/10.1016/j.spc.2025.01.022

Sgambaro, L., Chiaroni, D., & Urbinati, A. (2024). The design and servitization of products according to the circular economy principles: An ecosystem perspective in the building industry. *Journal of Cleaner Production*, 454, 142322. https://doi.org/10.1016/j.jclepro.2024.142322

Shedroff, N. (2009) *Design is the problem: The future of design must be sustainable.* New York: Rosenfeld Media

Simioni, F. J., Soares, J. F., Rosário, J.d. A.d., Sell, L. G., Bertol, E., Souza, F. M. P., Santos Júnior, E. P., & Coelho Junior, L. M. (2024). Industrial symbiosis and circular economy practices towards sustainability in forest-based clusters: Case studies in Southern Brazil. *Sustainability*, 16, 9258. https://doi.org/10.3390/su16219258

Yu, C., Fang, K., & Feng, Y. (2021). Circular economy in the construction industry: Industrial symbiosis pathways. *Journal of Cleaner Production*, 316, 128235. https://doi.org/10.1016/j.jclepro.2021.128235

2

Converting Thermoplastic Waste into Material Extrusion 3D Printed Parts: An Opportunity for Circular Economy

Sotiria Dimitrellou, Isidoros Iakovidis, and Eleni Strantzali

2.1 Introduction

Green transition and alignment with the three pillars of sustainability – i.e., economic, social and environmental – mandates the use of advanced materials and technologies that promote sustainable product development across various industries. Circular economy (CE) is a key driver for achieving sustainable development goals through the alignment with the R-strategies. R-strategies comprise a set of practices aimed at minimizing resource use, extending the life cycle of products and materials, and maximizing value recovery from waste. Recycling is a long-loop R-strategy which focuses on material exploitation and refers to the processing of waste into new products or materials that can be used for new products of the same or lower quality (Morseletto, 2020). The benefits resulting from adopting this strategy are both economic and environmental, including reduced dependency on natural resources, minimization of the volume of waste materials, and reduction of the environmental impact of the original feedstock material or product.

Material extrusion (MEX) additive manufacturing (AM) – also referred to as fused filament fabrication (FFF) or fused deposition modelling (FDM) – is a transforming three-dimensional (3D) printing technology, widely adopted for the manufacturing of commonly used or customized functional products for consumer or industrial applications. In MEX, a filament of thermoplastic material is extruded through a heated nozzle and deposited layer by layer on the 3D printer's build platform until the final object is complete. The contribution of MEX technology to Industry 4.0 is very important, as its operational principle, flexibility, and simplicity allows for the creation of complex and intricate geometries that cannot be produced by subtracting methods. The nature of the additive process combined with digital technology, including

DOI: 10.1201/9781003654889-2

computer-aided design and slicing software, leads to end-use products using the exact amount of material required, as defined in the CAD model, thereby minimizing material consumption and waste.

MEX 3D printing is compatible with a wide range of materials, including unreinforced thermoplastics, fiber-reinforced polymer composites, and ceramic and metal–polymer composites. As 3D printing is rapidly evolving into a competitive manufacturing technology, the use of more sustainable materials – such as bio-based, biodegradable, or recyclable materials – has emerged. By recycling post-industrial or post-consumer thermoplastic waste to produce new 3D printed products, the environmental impact of the original 3D printed material is reduced, and the AM process becomes even more sustainable, resulting in a lower carbon footprint. In a recent study (Yang et al., 2024), the health and environmental impacts of FDM 3D printing using a recycled polylactic acid (PLA) filament were compared to those of a virgin PLA filament. The results indicated that the use of the PLA recycled filament reduces the human health impact and carbon footprint of the 3D printing process and decreases resource wastage. However, PLA is a bio-based polymer derived from renewable sources and is biodegradable under specific composting conditions, which generally classifies it as an environmentally friendly material. Regarding waste that originates from fossil-based polymers which are difficult to degrade when discarded into the environment, recycling and remanufacturing (through 3D printing) becomes even more beneficial to promote environmental sustainability, financial viability, and achievement of CE targets (Morales et al., 2023; Punia and Kandasubramanian, 2025).

Most of the common thermoplastics for general use, as well as some engineering polymers, can be easily recycled due to their low melting point and viscosity. For high performance polymers (HPPs), recycling and reprocessing can be quite challenging, as HPPs exhibit high glass transition temperature and their molecular structure changes during processing. Although these materials are very promising for demanding high-strength applications and have also been utilized through 3D printing, very little research has been carried out to investigate their recyclability and their use as recycled filaments for the remanufacturing (3D printing) of components. Furthermore, to the authors' knowledge, no review article has been published so far which summarizes experimental studies regarding the recycling of HPPs for remanufacturing.

To address this issue of limited research and to enhance knowledge development, this chapter reviews current research on polymer recycling and component remanufacturing via 3D printing for a wide range of thermoplastics commonly used in engineering, including high performance polymers. Under this context, this review focuses on: (a) commodity polymers, such as acrylonitrile-butadiene-styrene copolymer (ABS), polypropylene (PP), and polylactic acid (PLA); (b) engineering polymers, such as polyethylene terephthalate glycol (PETG), polyamide (PA), polycarbonate (PC), polymethyl methacrylate (PMMA), and polyoxymethylene (POM); and (c) high performance polymers, such as polyvinylidene fluoride (PVDF), polyphenylene sulfide (PPS), polysulfone (PSU), polyether ether ketone (PEEK), polyetherketoneketone (PEKK), and polyetherimide (PEI).

This chapter contributes to the growing body of literature on this emerging topic, as follows.

- Overviews experimental studies on the recycling of HPPs and demonstrates the effect of the recycling process on the remanufactured components, with the aim to contribute to knowledge development and inspire future research.
- Highlights the challenges associated with the recycling and remanufacturing processes, and the influence of process parameters on the quality and properties of HPPs.
- Provides an overview of research studies on common and engineering recycled thermoplastics to allow for a comparative point of view, focusing on the effect of multiple extrusion and recycling cycles on the experimentally derived mechanical properties of 3D printed specimens.
- Indicates the main application areas for the polymers under consideration based on their suitability for the product requirements, highlighting the trend towards replacing metallic components with 3D printed HPPs for harsh industrial environments such as oil and gas, automotive, and aerospace.
- Provides future directions by proposing approaches that will provide systematic information and ensure experimental repeatability, as well as predict performance and optimize the AM process, aiming at the widespread adoption of recycled materials and circular manufacturing in industry.
- Potentially serves as a useful tool for academics, 3D printing practitioners, and industry professionals to visualize the opportunities that arise through the adoption of 3D printing and polymer recycling in the industry.

The rest of the chapter is organized as follows. In Section 2.2, thermoplastics used in industrial engineering sectors – such as oil and gas, automotive, and aerospace – are presented, and Section 2.3 summarizes the main challenges of polymer recycling. In Section 2.4, an overview of experimental studies on the characterization of recycled thermoplastic filaments and 3D printed parts focuses on commodity and engineering polymers. Common practices for the enhancement of recycled polymers through reinforcement and additives are described in Section 2.5. Section 2.6 highlights the challenges and state of the art literature on HPPs recycling, and Section 2.7 presents the conclusions and future research directions for HPPs recycling and circular manufacturing.

2.2 Thermoplastics for the Industrial Engineering Sectors

The replacement of metals with plastics is a common practice in the automotive and aerospace industries (Mishra and Jagadesh, 2022). Some of the advantages include weight reduction, corrosion resistance, noise and vibration damping, adaptability of design, innovative design, simplification of the production process (especially for complex geometries), and recyclability. Recently, there has been growing interest in replacing metals with polymer composites for the fabrication of functional components with stringent requirements both in small engines (Gurchinoff et al., 2013; Razavykia et al., 2020) and high-speed engines, such as turbines (Andrearczyk et al., 2021) and marine engines (Bertagna et al., 2022).

Traditionally, metallic or plastic-based components are produced by conventional manufacturing methods such as computer numerical control (CNC) machining, casting, or injection molding. However, the rapid development of MEX technology and recent advances in materials science allow the production of 3D printed functional components using fiber-reinforced polymer composites, metal-polymer composites, or HPPs that are suitable for demanding applications. Especially in cases when complex hollow geometry design, low volume production and cost constraints are required, additive manufacturing is a good substitute for conventional manufacturing for end-use products.

Thermoplastics that can be processed using MEX (FFF/FDM) technology are divided, as shown in Figure 2.1, into three categories, based on their thermomechanical properties: commodity, engineering, and high-performance polymers.

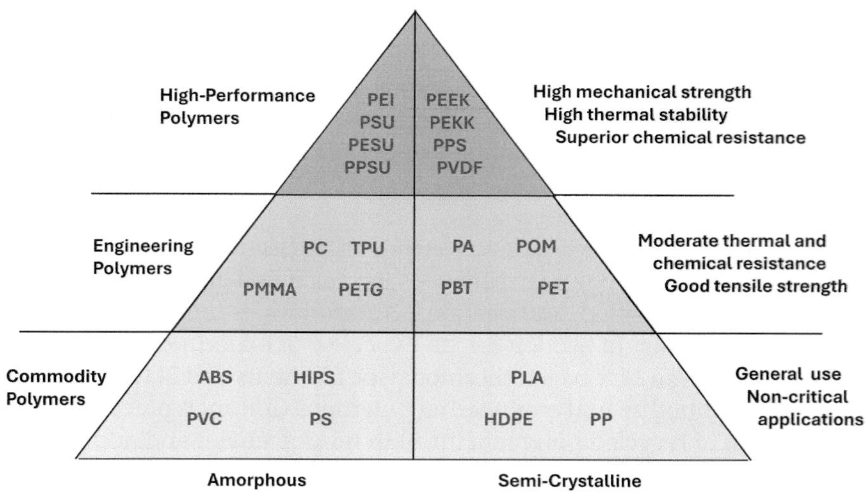

FIGURE 2.1
Hierarchy of thermoplastics commercially available as MEX filament.

In general, the most used thermoplastics in industrial and consumer applications, such as ABS, PP, PA, PC, and polyethylene terephthalate (PET), are cost-effective and lightweight materials, with low to moderate thermomechanical properties, suitable for applications with no strict requirements. In the automotive industry, common fossil-based polymers used for the fabrication of instrument panels, fuel systems, lighting systems, body parts, interior and exterior trim, and bumpers, as well as other components, include ABS, PP, PC, PA, polystyrene (PS) and polybutylene terephthalate (PBT) (Vieyra et al., 2022).

For more demanding engineering applications, the properties of thermoplastics have been enhanced using reinforcement such as carbon, glass, and aramid fibers. Fiber-reinforced polymer composites have a high strength-to-weight ratio, good anti-wear properties, and improved anti-aging behavior compared to traditional metals (Pervaiz et al., 2021). For example, the addition of carbon fibers to the polypropylene matrix (PP-CF) has been found to improve the tribological behavior of 3D printed samples and is recommended for upgrading sliding or rotating components in the automobile industry (Karthick et al., 2024). PP-CF filament is also ideal for lightweight marine equipment and automotive mechanical parts, due to polypropylene's near-zero water absorption rate and its resistance to chemicals and UV radiation. In the work of Bertagna (Bertagna et al., 2022), glass fiber-reinforced polyamide composites (PA-GF) were examined in terms of performance and cost for a camshaft cover of a four-stroke marine engine. It was observed that the most suitable material was PA6.6/30%GF, which resulted in a 60% weight decrease and 50% cost decrease of the original aluminum-alloy cover.

POM and PMMA engineering polymers have recently received attention for 3D printing applications due to their special characteristics. POM exhibits high wear resistance and low friction properties, and is commonly used for the manufacturing of bearing surfaces, pulley wheels, housing parts, and gears (Zhang et al., 2021). PMMA is an optically transparent thermoplastic which exhibits high electrical resistivity and weathering resistance, with wide application in vehicles, electronic devices, biomedical, and energy storage (Vidakis et al., 2023).

HPPs are appropriate for high-temperature industrial applications that require components with enhanced strength and resistance to the chemical environment – i.e., acids, fuel mixtures, or mineral engine oils – that can cause corrosion to metals. In addition, HPPs can retain their properties when subjected to high temperatures and exhibit superior friction and wear performance, and thus can be used in seal rings, thrust washers, bushings, valve caps, and possibly piston rings (Gurchinoff et al., 2013). The potential use of PVDF, PPS, PSU, polyphenylsulfone (PPSU), PEEK, and PEI in the oil and gas industry and their physical and thermomechanical properties have been presented in de Leon et al. (2021). The tribological characteristics of HPPs and their composites – including wear and friction in sliding when exposed to high temperature, vibration, and lubrication conditions – have also been explored and discussed against their metallic counterparts (Kurdi and Chang, 2019). The resistance of FFF-printed PVDF, PPSU, PEI, and PEEK to harsh environmental conditions has been examined in terms of oil compatibility, dielectric strength, and kerosene compatibility (Costa et al., 2024). FFF-printed PPS specimens have shown acceptable performance when immersed in automobile fluids, i.e., car coolant, diesel, and gasoline, indicating their potential use in fluid transfer automotive parts (Garmabi et al., 2022). Similarly, the properties of additively manufactured PEI have been investigated and its potential use in automotive applications (Nguyen et al., 2022) and the aerospace industry (Boado-Cuartero et al., 2024) has been highlighted. PEEK and PEKK belong to the polyaryletherketone (PAEK) family, with superior chemical and thermomechanical properties suitable for extreme environmental conditions, such as those in the oil and gas industry (Geslin et al., 2021) and in aerospace technology (de Seijas et al., 2024).

The growing interest in replacing metal components with corresponding polymers is expected to lead to further increases in plastic waste. Post-consumer plastic waste can originate from equipment and machinery that have limited life span and require regular maintenance and repair activities, such as gears, bushings, clevis joints, bearings, pillow block units, housings, linear guides, and other engine components. Post-industrial plastic waste can include scraps, cut-offs, and discarded materials generated during the manufacturing process. In addition, the promising characteristics of advanced high-strength thermoplastics – including fiber-reinforced polymers, metal-polymer composites, or HPPs – are expected to increase their demand in harsh industrial environments, inevitably leading to increased plastic waste production. Following this trend, the need to integrate recycling technologies with advanced manufacturing processes, to align with product sustainability, becomes even more urgent. Figure 2.2 illustrates the transition from linear economy to circular economy through the integration of MEX 3D printing and plastic waste recycling. By adopting this CE model, the concerned industries could reduce the amount of materials required for manufacturing, maximize resource efficiency, and become sustainable and cost-efficient.

A practical direction for small-scale mechanical recycling could be the exploitation of the operational flexibility of the FFF/FDM technology that allows for in-situ recycling and in-situ manufacturing, so that new products can be additively manufactured directly from the generated waste. In such a case, waste can originate entirely from discarded 3D printed components or scrap – such as misprints, supports, and defective filament – and the recycling process can be straightforward and more efficient in terms of sorting, in-house facilities, and reprocessing. Closed-loop recycling can be easily employed for small-scale manufacturers and laboratories, as well as remote facilities including marine vessels, offshore platform rigs, and space stations.

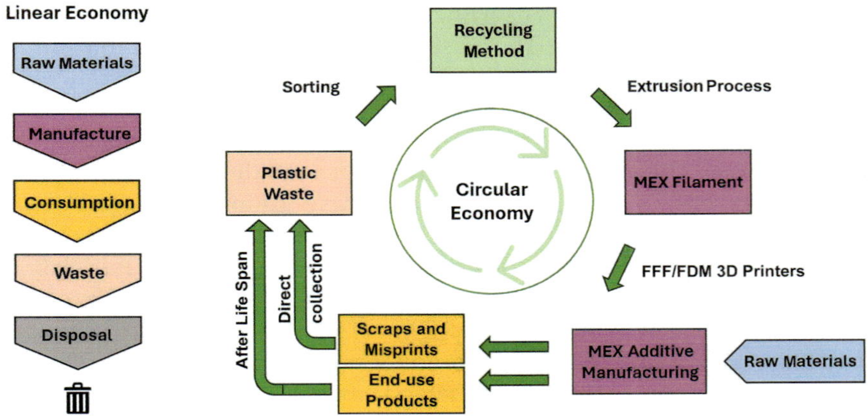

FIGURE 2.2
The shift from linear economy to circular economy, with conversion of plastic waste into MEX 3D printed parts.

2.3 The Challenges of Processing Materials Through Recycling

Different recycling methods for polymers have been presented in the literature, including mechanical, chemical, and thermal recycling. Thermal recycling uses processes such as pyrolysis and fluidized bed combustion to heat polymer to elevated temperatures and decompose it into smaller molecules (Punia and Kandasubramanian, 2025; Türker et al., 2024). The primary challenge is the energy intensity demand and the degradation of the quality of the recycled polymer which is affected by the thermal degradation process (Punia and Kandasubramanian, 2025; Rashid and Koç, 2023). In chemical recycling, chemical agents are used to break down polymer into constituent monomers and other chemical constituents. This process allows for the recycling of materials that are difficult to recycle using other methods, such as fiber-reinforced polymers, as it preserves the fiber characteristics and length (Türker et al., 2024). However, chemical recycling presents challenges including the use of hazardous chemicals and the generation of impurities (Türker et al., 2024; Rashid and Koç, 2023). Mechanical recycling is a popular and cost-effective method which does not significantly affect the chemical composition or destroy the molecular structures of the polymers, resulting in little or no deterioration in quality (Punia and Kandasubramanian, 2025; Türker et al., 2024). All types of thermoplastics can undergo mechanical recycling, but contaminated waste or fiber-reinforced polymers are not recovered efficiently by this method (Türker et al., 2024). The mechanical recycling process involves several steps such as sorting, purification, grinding, melting, and reprocessing of the polymer. The waste is first sorted/separated into similar materials to ensure that the recycled polymer is of consistent quality. The sorted batch of material is then ground up or shredded into granules, flakes, or pellets, and then melted. Depending on the manufacturing method, the melted polymer can either be used directly for molding or converted into filaments via screw extrusion for 3D printing applications.

According to several studies in the literature (Morales et al., 2023; Mishra et al., 2023a), major considerations that affect the efficiency of recyclability include: (a) the presence of additives and coatings that are difficult to separate from the base polymer due to polymers' low molecular diffusion rate; (b) the aging by weathering of the original material in the course of its life that reduces the percentage of the material that can be recycled; and (c) the impact of the recycling and extrusion processes on the physical and mechanical properties of the material, resulting in degradation of quality.

In most experimental studies, the recycled filament is either purchased by a supplier or produced via the recycling and extrusion process (Figure 2.3). The latter includes the grinding of the plastic waste, the creation of granules or pellets, their melting, and the extrusion of the melted polymer into a filament using a rotating shaft and a screw extruder. During this process, user-defined parameters such as melting temperature, shaft spindle speed, fan cooling, extrusion temperature, extrusion speed, and nozzle diameter affect the recycled MEX filament quality. As the material waste is melted and then exposed to shear stress through the nozzle extruder, its interlayer bonding strength is weakened (Morales et al., 2023; Mishra et al., 2023a). This results, in most cases, in slightly lower mechanical performance of the recycled filament, compared to that of its original counterpart (i.e., virgin material).

Regarding the remanufacturing (3D printing) of new parts using the recycled filament, the mechanical behavior and the quality of the 3D printed

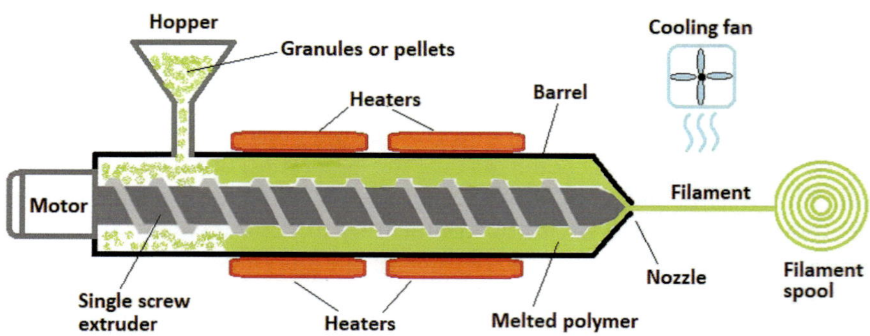

FIGURE 2.3
Schematic representation of converting polymer pellets to recycled filament using a screw extruder.

parts depend significantly on the recycled filament quality and the MEX printing parameters. Specific printing parameters, such as nozzle diameter, nozzle temperature, print speed, and bed temperature, must be set according to the recycled filament's physical properties, i.e., glass transition temperature, melt volume rate, and thermal shrinkage. Other printing parameters, such as layer height, infill density, infill pattern, build orientation, and raster angle can differentiate according to the desired properties for the final part. However, slight variations in the values of the printing parameters affect directly the performance of the 3D printed part, including strength, dimensional accuracy, thermal stability, and printability. Printability can be defined as the ability of a material to maintain the desired print outcomes for the specific application. When assessing the printability of a material, consideration should be given to the extrudability, filament deposition, and printing accuracy of the intended shape and size, as well as reproducibility and repeatability of the part dimensions when the same printing parameters are used (Dimitrellou et al., 2025).

Therefore, it can be concluded that a complex combination of process parameters affects the physical, rheological, and mechanical properties of the recycled filaments and the consequent remanufactured components. The process parameters are not standardized, but are determined according to the filament supplier guidelines, process parameters reported in literature for the same material, or trial and error testing. Thus, the properties of the final product are not guaranteed, even when the same original material is recycled. This clearly indicates the importance of carrying out several studies on the characterization of recycled thermoplastics to ensure experimental reproducibility and verify the viability of recycled polymers.

2.4 Exploring the Mechanical Behavior of Recycled Thermoplastic Filaments

Most of the commodity thermoplastics – such as PLA, ABS, PET, and PP – can be easily recycled due to their low melting point and viscosity, and experimental findings relative to 3D printed samples of recycled filaments have been summarized in review articles in the literature (Mishra et al., 2023a; Shanmugam et al., 2020). For engineering polymers, such as PC, PA, PMMA, and POM, there is a very limited number of studies that refer to the processing of waste into recycled filament for the fabrication of new 3D printed products. An overview of the most representative studies on the experimental investigation of recycled commodity and engineering thermoplastics for 3D printing application follows.

Recycled PLA is biodegradable, easily processable, and widely commercially available as MEX printing filament. Repeated recycling has been linked with a decrease in viscosity, which in turn reduces tensile strength and stiffness, suggesting that repeated recycling of PLA may not yield a reliable product. Cruz et al. (2017) investigated the properties of recycled PLA 3D printing samples up to five recycling cycles. The results showed an important reduction of molecular weights after multiple cycles (26.73% after three cycles and 46.91% after five cycles), a reduction in viscosity, and an important increase in the melt flow index of about 6.05 times. The tensile strength and elastic modulus of the recycled PLA samples after one cycle were comparable to virgin PLA; however, increasing the number of recycling cycles resulted in the degradation of these properties.

The investigation of the thermal, rheological, and mechanical properties of virgin PLA pellets and recycled PLA feedstock up to six extrusion and recycling cycles for large-format AM systems has been presented (Romani et al., 2024). It was observed that thermal properties exhibited slight variations after multiple cycles and that brittleness was increased by repeated extrusion and recycling cycles, while good printability was achieved. The elastic modulus was increased by 6% after the first and second recycling cycles, while after six extrusion cycles, its value corresponded to a minimal decrease (–1.0%) compared to the value of the first extrusion cycle (virgin pellet). The ultimate tensile strength showed a minor variation after the first two recycling cycles, while after the third recycling cycle, it was decreased by 8.1%. The elongation at break (%) showed small percentage reductions for the first three recycling cycles with a significant decrease after the fourth cycle (–30.1%).

In a recent study (Hasan et al., 2024b), the optimum FDM printing parameters for virgin PLA (vPLA) and recycled PLA (rPLA) filaments were examined. The experiments showed that for the same printing parameters, the rPLA samples exhibited lower tensile strength, flexural strength, and hardness, and similar impact strength. In addition, surface roughness was increased, mainly due to the lower molecular weight caused by the mechanical recycling process and the consequent lower viscosity of the recycled material. However, using the user-defined optimum printing parameters, rPLA showed a high tensile strength of 45.97 MPa, which was comparable to that of vPLA at 80% infill density.

Nagengast et al. (2023) investigated the properties of one stage mechanically recycled PLA, PP, and PP support filaments. First, original 3D printed specimens that had been simulated by an accelerated aging process similar to that of prototypes or short-life span parts were used for the recycling and extrusion process. The experiments showed that the degree of crystallization and the melt flow rate (MFR) were approximately the same for virgin and recycled PP materials. On the contrary, the MFR of the PLA filament significantly increased with the recycling process. Regarding the mechanical performance, decreases of 11.75% in E-modulus and 17.5% in tensile stress were observed for rPLA samples. For PP samples, no detectable degradation of the main properties occurred during recycling. E-modulus and tensile stress increased by 3.4% and 15.4%, respectively for the recycled PP, while for recycled PP support, an increase of 1.7% and 6.25% was observed, respectively. Also, the average surface roughness was increased for rPLA (35.1%), while a very small increase of 1% was observed for rPP samples. All results indicated that recycled PP has acceptable performance and has strong potential to be used for additively remanufactured parts.

In another experimental study (Dobrzańska-Danikiewicz et al., 2023), the tensile strength and micro-hardness of recycled PLA and recycled PETG samples were found to be similar to virgin PLA and virgin PETG for one recycling round. Small variations in these values were noted by changing the printing temperature and filament deposition. Another study in this context (Strano et al., 2023) showed that rPETG and rPLA samples originated from post-industrial waste exhibited good printability, while tensile and flexural strength were slightly higher compared to virgin materials. The energy consumption required to produce virgin and recycled PETG and PLA (including pellet production and filament extrusion) was also calculated, concluding that recycling allows for significant energy savings of more than 50%.

Tensile, flexural, and fracture toughness and micro hardness tests have also been conducted (Vidakis et al., 2021b) for multiple recycling cycles for PETG. The results revealed that the overall mechanical strength was slightly improved up to the third and fourth rounds, while a minor degradation effect on the thermomechanical properties of PETG after the fifth round of recycling. The maximum values obtained for the investigated properties are shown in color in the table in Figure 2.4. In the third recycling round, a

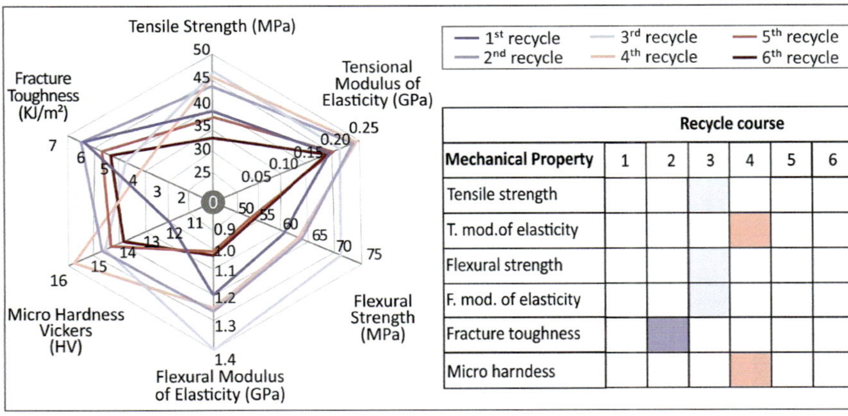

Mechanical Property	1	2	3	4	5	6
			Recycle course			
Tensile strength						
T. mod.of elasticity						
Flexural strength						
F. mod. of elasticity						
Fracture toughness						
Micro harndess						

FIGURE 2.4
Overall results on the mechanical properties of virgin and recycled PETG in the six rounds of recycling.

Source: Vidakis et al. (2021b)

maximum value was observed for tensile and flexural strength, corresponding to an increase of 15.8% and 17.8%, respectively, compared to virgin PETG. Tensile modulus of elasticity and microhardness also showed maximum values in the fourth recycling round. The printability was also satisfactory up to the fourth round due to the high melt flow rate of the recycled PETG.

Regarding ABS, most literature studies have reported that the first reprocessing cycles of ABS lead to good control of the melt flow rate and minimal difference in tensile properties. Strano et al. (2023) found that samples made from recycled rABS exhibited similar printability and a slight decrease in tensile strength, stiffness, and ductility compared to virgin ABS. In addition, when exposed to 400 hours of ultraviolet (UV) light irradiation, rABS samples exhibited similar tensile strength, slightly reduced ductility and printability, and higher stiffness. Cress et al. (2021) investigated the effect of repeated recycling on the mechanical behavior, molecular structure, and chemical composition of ABS FDM-printed samples. The experimental tests showed no significant change in the molecular structure after each round of recycling; nevertheless, porosity was increased, resulting in the degradation of mechanical properties. Specifically, during the third round of recycling, tensile and fracture strengths were reduced by 10%, strain at break by 25%, and toughness by 37%.

According to Vidakis et al. (2021a), PA12 was successfully recycled for a certain number of cycles and FFF-printed samples were found to exhibit improved mechanical properties between the second and third recycling course compared to virgin material. Figure 2.5 presents the average values of

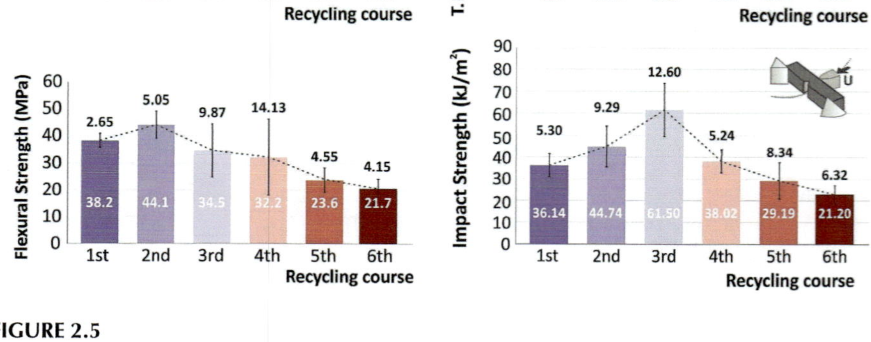

FIGURE 2.5
Comparison of mechanical properties' average values of recycled PA12, along with the calculated deviation for all six recycling courses.

Source: Vidakis et al. (2021a)

tensile strength, modulus of elasticity, and flexural and impact strengths for six recycling courses. The results showed a peak value for tensile strength at the fourth course, which corresponds to an increase of 17.7% when compared to virgin PA12, and a significant improvement of impact strength at the third course, which corresponds to an increase of 47.3% compared to the first recycling course. In addition, thermal analysis indicated a slight decrease in the crystallinity degree after the fourth recycling course, while morphology characterization of the samples' side surfaces showed good printability and uniform layer deposition without gaps and discontinuities for the first to fourth recycling courses.

Reich et al. (2019) performed a comparison of the mechanical properties of specimens fabricated from recycled PC regrinds and virgin PC filament using powder melt extrusion (PME)-based and FFF-based 3D printers, respectively. It was observed that the tensile strength of recycled PC was comparable to that of the virgin material, while the compressive strength was found to be lower. In addition, the large nozzle size of the PME-based 3D printer resulted in an increased volumetric deposition rate of the PC regrinds, reduced printing time, and increased ease of printing for large objects. In another recent study (Costa et al., 2023), samples of virgin and recycled PC were produced by injection molding and their thermal and mechanical properties compared. The experimental tests revealed a slight decrease in density and glass transition temperature after each reprocessing cycle, a small increase in the melt flow rate and hardness, and a minimal decrease in tensile strength and Young's modulus for the recycled PC samples.

Regarding PMMA, the use of recycled PMMA sheets to produce sensor systems (Prado et al., 2017) and the possibility of recycling PMMA through chemical processes (Sponchioni and Altinok, 2022) have been explored. In the work of Vidakis et al. (2025), PMMA sheet trimmings were shredded into pellets and then melt-based extruded to produce filaments for 3D printing of specimens. The specimens were shredded again into pellets, and the entire process was repeated for six successive cycles. The results were very promising, as PMMA showed thermomechanical stability, marginal thermal degradation, and persistent printability throughout the mechanical recycling process, while tensile and flexural mechanical properties were enhanced in the second cycle.

While POM is extensively used in the electronics, construction, and automotive sectors, no study has been published on the remanufacturing of samples through MEX 3D printing using recycled POM filament. However, samples have been manufactured by injection molding using pure POM regrinds which were mechanically recycled using post-industrial waste, and it was found that they resulted in a lower product carbon footprint than samples manufactured using virgin POM (Tinz et al., 2022). It has also been reported that POM is currently only marginally recycled as the extrusion or regranulation processes of POM when mixed in other plastic waste streams can lead to the formation of toxic formaldehyde vapors (Olscher et al., 2022).

2.5 Enhancing the Performance of Recycled Thermoplastics

The main challenge regarding filament production from polymer waste is the degradation of the properties of the original material which occurs during the recycling process. It has been reported in the literature that polymers undergo chain scission when subjected to thermomechanical recycling, reducing the molecular weight and weakening the interlayer bonding (Mishra et al., 2023a). During the recycling and extrusion processes, the polymer is exposed to shear stress, oxygen, and high temperature, which have significant impact on the molecular weight and several physical properties, such as viscosity, porosity, degree of crystallinity, and thermal stability (Morales et al., 2023). These changes in morphological characteristics cause degradation of the mechanical properties – i.e., Young's modulus, tensile strength, and toughness – and deteriorate the surface finish and printing quality.

Generally, the quality and performance of the recycled filament can be improved by adding suitable reinforcements and blends that will assist in enhancing mechanical, rheological, and thermal properties. A simple and cost-effective approach is the blending of virgin material with recycled material in various weight proportions, which can increase molecular weight and improve mechanical properties. Another approach is to reinforce the recycled polymer with ceramics, natural or synthetic fibers, or other fillers (e.g. iron powder or graphene). The addition of fibers and fillers at an optimum percentage ensures proper flow velocity and provides good print quality and dimensional stability of the 3D printed samples (Lodha et al., 2023).

Comprehensive reviews of common recycled polymers regarding their behavior after the use of reinforcements and blends have been presented, including PLA, ABS, high-density polyethylene (HDPE), low-density polyethylene (LDPE), and PA12 (Shanmugam et al., 2020); PLA, PP, PA, ABS, PET and HDPE (Lodha et al., 2023); and PLA, ABS, HDPE, LDPE, PP and PET (Mishra et al., 2023a). Regarding engineering thermoplastics, such as PC and PA, a small number of studies have been conducted, indicating the need for additional experimental data to ensure the quality and performance characteristics. An overview of representative studies on the recycling of commodity and engineering thermoplastics combined with additives and their reprocessing to MEX printed parts is presented in what follows.

In a study conducted by Tanney et al. (2017), the impact of blending virgin PLA pellets with recycled PLA at different ratios was examined for two recycling phases. For the first recycling phase, the addition of 75% vPLA in the blend resulted in an improved tensile strength compared to pure rPLA. For the second recycling phase, the average ultimate tensile strength (UTS) of the measured samples was decreased by approximately 22% compared to that of vPLA. The standard error of UTS values for the second recycling phase was also higher when compared to the first phase, indicating that the recycled filament loses its consistency during recycling, possibly due to small variations in the filament's extrusion diameter.

Hasan et al. (2024a) investigated the tensile, flexural, and impact properties, the surface roughness, and the hardness of 3D printed samples of virgin PLA (vPLA), recycled PLA (rPLA), and blends of 50% virgin and 50% post-industrial recycled waste. The results revealed that blending improved the tensile, flexural, and impact strengths and hardness by 18.40%, 8.00%, 9.15% and 5.00%, respectively, compared to those of rPLA. Also, a significant decrease in surface roughness of 2.29% compared to rPLA was obtained, due to the improved melt flow and cooling behavior of the added vPLA, while blended samples exhibited better interlayer adhesion, as evidenced by the fractographic analysis.

In another study (Pickering and Stoof, 2017), sustainable 3D printing filaments were manufactured by reinforcing post-consumer recycled PP (rPP) with varying weight fractions (10%, 20%, and 30%) of hemp or harakeke fiber. It was found that the 30 wt% harakeke–rPP filament exhibited a significant increase in tensile strength and Young's modulus of 77% and 275% respectively, relative to the unreinforced rPP filament. In addition, the reinforcement with natural fibers was found to reduce the shrinkage of the 3D printed samples, with a maximum reduction of 84% for the 30% harakeke recycled PP composite. Recent experimental studies have revealed that the reinforcement of virgin and recycled PP matrices with synthetic fibers – i.e. glass (Sam-Daliri et al., 2023) or carbon (Zhan et al., 2024) – is promising for the enhancement of the mechanical properties of the 3D printed specimens and the reduction of the warping effect of virgin PP during 3D printing.

Recycled ABS has been reported to exhibit poor printability, including shrinkage and low bed adhesion. In this context, blending recycled ABS with virgin ABS has been investigated for the improvement of the inter-layer bonding and its effect on the performance of the recycled material (Mishra et al., 2023b). In the same study, the properties of 3D printed samples produced by blending virgin ABS with recycled ABS in varying weight proportions and different printing temperatures were explored. The results indicate that some samples exhibited higher mechanical properties, e.g. the 60%rABS/40%vABS sample printed at 240°C showed an increase in Young's modulus, yield strength, and ultimate tensile strength of 13.69%, 13.37%, and 15.64%, respectively. Figure 2.6a reveals that blends with higher vABS content (30% to 50%) exhibited higher melt flow index (MFI), improved material fluidity and good printability, mainly due to an increased molecular weight compared to rABS. In addition, for blends with vABS content greater than 30%, the consistency in melt flow and cooling behavior resulted in improved surface finish. The lowest surface roughness of 1.027μm was obtained for the 60% rABS/40% vABS sample printed at 230°C, which corresponds to a decrease of 38.39% compared to vABS (Figure 2.6b).

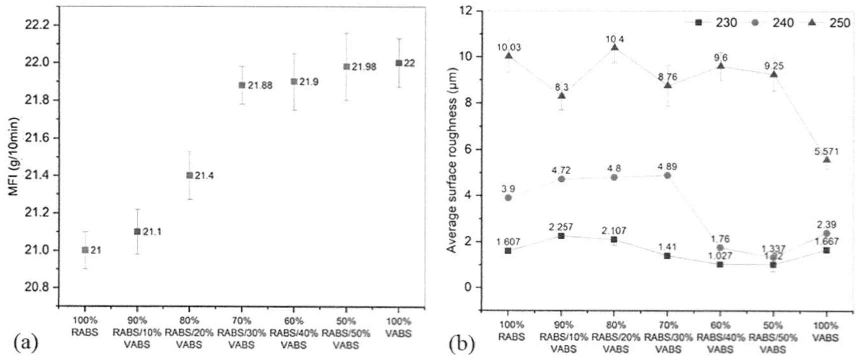

FIGURE 2.6
Melt flow index (a) and average surface roughness (b) of different rABS/vABS samples.

Source: Mishra et al. (2023b)

A review on the recycling of ABS waste and its blending with other standard and engineering polymers – such as PP, PC, polyvinyl chloride (PVC), and high impact polystyrene (HIPS) – has been presented in (Deshmukh et al., 2024). According to the same paper, the blending of rPC with rABS containing a polyethylene/poly(glycidyl-methacrylate) copolymer as compatibilizer and solid epoxy resulted in improved mechanical properties due to the successful combination of the high processability of ABS and the high heat distortion temperature of PC. In a similar case, blends of PC/ABS and PA6/PC/ABS were recycled five times and the properties of specimens that were produced by injection molding were investigated (Kumar et al., 2015). The rheology of the specimens was also evaluated via measuring melt flow index. The results revealed that the thermal properties and chemical composition of both blends did not significantly change with recycling. For the PC/ABS blends, recycling slightly affected the tensile and flexural strengths; however, the melt flow index and impact strength were decreased compared to the virgin counterparts. For the PA6/PC/ABS blends, the elastic modulus and tensile strength were decreased with the number of recycling stages, while the flexural and impact strengths and the melt flow index were improved.

To exploit the benefits of using natural fibers, the properties of FDM-printed specimens of recycled ABS and PA6 reinforced with banana fibers have been investigated (Singh et al., 2019). In the case of PA6, the reinforcement was found to reduce the porosity of the surface and to improve several mechanical properties, including the load resistance capacity, tensile strength, elongation, and Young's modulus. In the case of recycled ABS, the reinforcement with banana fibers resulted in the improvement of load resistance capacity and tensile strength, whereas elongation at break and Young's modulus were decreased.

The reinforcement of polyamide with glass or carbon fibers is also a successful practice to enhance its properties for demanding engineering applications. In Al-Mazrouei et al. (2022), samples of recycled nylon composites were processed using a twin-screw extruder and a compression molding process. The blends were produced by mixing two 3D-printed composite wastes – i.e., carbon fiber/nylon and glass fiber/nylon – at different mixing ratios. It was observed that for specific samples higher values of tensile strength, Young's modulus, ductility, and toughness were attained compared to pure composites. In a recent study (Su et al., 2024), PA6 discarder fishnet (rPA) and carbon fibers waste (rCF) were reclaimed, mixed, ground up, and used as raw material in a single extruder for filament production. Specimens were produced using an FDM-based printer and tensile tests were conducted with different printing parameters. It was observed that the tensile strength and modulus of the best performance specimen were 76.5 MPa and 4.8 MPa, respectively, indicating a strong potential for in-house recycling and fabrication of rPA–rCF composites. In a similar context, samples of recycled PA66 reinforced with glass fibers, which are widely used in the automotive industry, were produced by injection molding (Gültürk and Berber, 2023). The mechanical tests showed an increase in melt flow index and elongation at break, while tensile strength, elastic modulus, and impact energy were slightly decreased.

2.6 The Potential of Using Recycled HPP Filament for MEX 3D Printing

2.6.1 Challenges in HPP Filament Processing

The superior performance of HPPs has increased their demand, especially in industrial sectors whereby resistance to high loads, chemicals, and high temperature conditions is required. According to Weyhrich and Long (2022), HPPs display thermal decomposition temperatures (T_d 5% weight loss) > 450°C, long-term durability (>10,000 h) at 177°C, and a high glass transition temperature (T_g > 200°C) due to compositions containing rigid aromatic repeat units. However, their processing through additive manufacturing technology – such as fused deposition fabrication, selective laser sintering technology, or laser powder bed fusion – can be challenging due to their high T_g values and melt viscosities.

The basic challenges arise from the chemical complexities of these polymers of highly thermal stable bonds and the effect of the fabrication process on the molecular structure and thermal and rheological properties (Das et al., 2020). As processing occurs above the polymer's glass transition temperature, long processing times have the potential to alter molecular structure, thereby changing the properties of the fabricated part. Moreover, semi-crystalline high temperature polymers – such as PEEK, PPS, and PVDF exhibit high shrinkage coefficient; thus, MEX 3D printing is challenging as volume shrinkage during printing reduces the dimensional and geometrical accuracy of the printed parts. On the other hand, high crystallinity restricts the motion of the polymer chains, which allows semi-crystalline HPPs to maintain their mechanical properties under long exposure in higher temperatures (Das et al., 2020). Regarding amorphous HPPs, such as PEI, PSU, and PPSU, they have a more rigid molecular structure and interlayer bonding. These characteristics can reduce residual stress during filament deposition and cooling, resulting in better printability, no wrapping, and good dimensional and geometrical accuracy.

Although some HPPs – such as PPS, PPSU, PEI, and PEEK – can be recycled to regrinds or powders for use in injection molding, compression molding, or composites (e.g., PolyClean Technologies, Inc. is a company specialized in high temperature polymer recycling), whereas only PEEK has been commercially available as recycled MEX 3D printing filament.

Extensive literature search revealed that a limited number of experimental studies have investigated the properties of specimens manufactured either by injection molding or 3D printing using recycled HPPs. In the majority of these studies, the specimens have been produced by injection molding using recycled HPP regrind or pellets. However, injection molding does not significantly affect the molecular structure of the material or its rheological properties. On the contrary, in the case of the screw extrusion process, the quality and performance of the recycled HPP filaments are directly dependent on the process parameters (melt temperature, shaft spindle speed, extrusion temperature, melt strand diameter, etc.), as mentioned in Section 2.3. Accordingly, the quality and properties of the remanufactured (3D-printed) specimens depend on: (a) the quality and properties of the recycled HPP filament; and (b) the conditions of the subsequent MEX printing process (high nozzle temperature, maintainable temperature in the chamber, print speed, layer height, raster angle, raster width, etc.).

2.6.2 Recycling of PVDF

In the recent work of Mehta et al. (2024), filament samples of recycled PVDF and recycled PVDF composites were developed using a twin-screw extruder with three repetitions and different extrusion parameters. For the composites, additives commonly used in conventional dry cells – such as MnO_2, graphite, $ZnCl_2$, and NH_4Cl – were incorporated into the rPVDF matrix in varying proportions. Tensile tests were conducted for all filaments and the filament of each batch with the highest Young's modulus was selected to test the printability. However, unsuccessful 3D printing was observed for the produced composites due to the elastic nature of PVDF, so 40% LDPE was added into the rPVDF matrix to improve stiffness and printability. Experiments also revealed that for specific samples, mechanical properties such as Young's modulus, tensile strength at peak and strength at break were increased significantly compared to virgin PVDF.

In a similar study (Husain et al., 2023), filament samples of recycled PVDF composites were produced using a single screw extruder and was compared with filaments of virgin PVDF. The experimental tests revealed a significant improvement of 122.6% in viscosity, an increase of 45% in Young's modulus, a slight variation in thermal and dielectric properties, a slight increase of 6.79% in surface roughness, and a significant decrease of 88.37% in toughness modulus. The results indicated that recycled PVDF has the potential to be used for 3D printed functional sensors or other prototypes which are not operating in crash-worthiness applications.

2.6.3 Recycling of Polysulfones

PSU, PPSU, and polyethersulfone (PESU) belong to polysulfones and are amorphous high-temperature thermoplastics with high chemical resistance and high mechanical strength suitable for a wide range of applications (Maletzko and Herz, 2011). Although PPSU and PESU are more widely available as commercial FDM printing filaments than PSU, the recycling of PPSU or PESU waste for MEX filament extrusion has so far – to the knowledge of the authors – not been investigated in the literature. A limited number of studies on the characterization of injection molded samples from recycled PSU are presented in what follows.

In the study of Yu et al. (2019), the mechanical properties, thermal stability, and long-term durability of virgin PSU and recycled PSU were investigated. Wasted nonwovens were collected, melted, and converted into PSU pellets through a two-stage process, using an extruder machine. Virgin PSU (vP) and recycled PSU pellets (rP1 and rP2) were then injection molded into dumbbell-shaped specimens. The mechanical tests revealed no difference in tensile strength between the virgin and recycled PSU samples, as depicted in Figure 2.7a, indicating an excellent degradation resistance. A slight decrease in elongation was observed for the recycled PSU due to the impurities produced during the recycling process. In addition, the thermal stability of virgin PSU was found to be superior due to the degradation of the recycled material. However, a drastic decrease in impact strength of 28.71% and 41.30% was observed for the first-time and second-time recycled PSU, respectively, as shown in Figure 2.7b.

Li et al. (2022) expanded on these findings and investigated the hydrothermal aging resistance of virgin PSU and recycled PSU by conducting tensile, three-point bending, impact strength, fracture toughness, stress relaxation, and creep tests. The samples were fabricated by injection molding and were immersed in 98°C hot water for one year. The results revealed that the virgin PSU samples exhibited excellent resistance to long-term hydrothermal aging, while the strength (tensile, flexural, and impact) and toughness of the recycled PSU samples were found to decrease from the ninth to the 12th month.

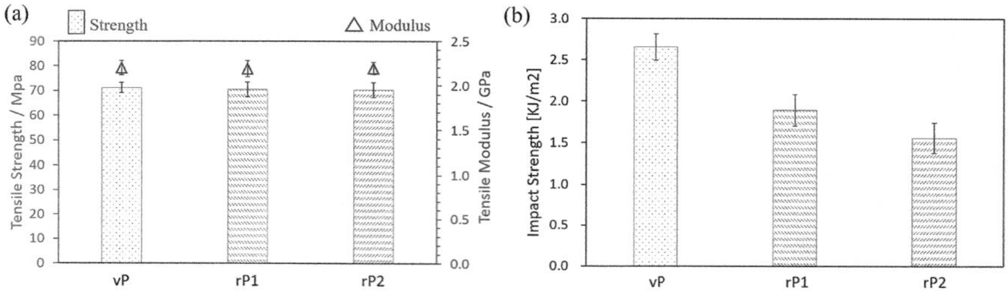

FIGURE 2.7
Tensile strength (a) and impact strength (b) for vP (virgin PSU), rP1 (first-time recycled PSU), and rP2 (second-time recycled PSU).

Source: Yu et al. (2019)

2.6.4 Recycling of PPS

A review of viable recycling processes for PPS-based waste – including carbon and glass fiber-reinforced PPS composites, which are widely used in automotive and aerospace industries – is presented by Montagna et al. (2022). According to the authors, the most common process for recycling PPS is mechanical recycling, while chemical recycling is challenging due to its chemical inertness and insolubility in organic solvents.

In a recent study (Dollischek et al., 2024), samples of post-industrial recycled glass fiber–reinforced PPS composites were produced by injection molding and the economic benefits of recycling on an industrial scale were highlighted. The experimental tests showed that the recycled PPS-GF samples exhibited higher melt flow index, lower viscosity, higher dimensional shrinkage, and lower tensile strength relative to the virgin material. Furthermore, it was found that recycling PPS-GF can lead to high cost savings (up to 44%) without a significant reduction in performance.

The mechanical recycling of PPS continuous fiber self-reinforced composites (CFSRCs) and their advantageous use in space applications was presented in Zhang et al. (2023). First, 3D printed composites were recycled and converted into 3D printing filament by screw extrusion. The experimental tests on the new 3D printed samples indicated that the tensile properties of the recycled PPS CFSRCs samples were lower than those of pure PPS CFSRCs, but higher than those of the virgin PPS matrix, as depicted in Figure 2.8a. A similar improvement was also observed for the flexural and impact strength. The improvement in mechanical properties was further attributed to the incorporation of the PPS fibers in the composites, which – although melted during the recycling process – still possess oriented macromolecular chains, serving as a reinforcement. Furthermore, it was found that the number of recycling cycles for the PPS CFSRCs samples slightly affected the tensile strength, and higher values were obtained at the fourth and fifth cycles, as shown in Figure 2.8b.

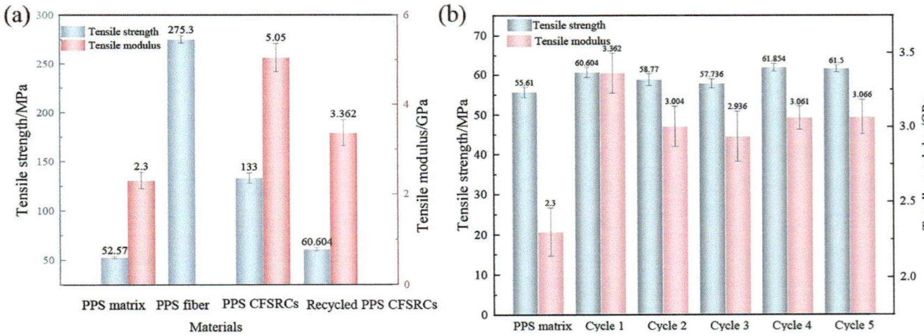

FIGURE 2.8
Tensile properties of recycled PPS CFSRCs compared with materials (a) and ultimate tensile properties of recycled PPS CFSRCs for multiple cycles (b).

Source: Zhang et al. (2023)

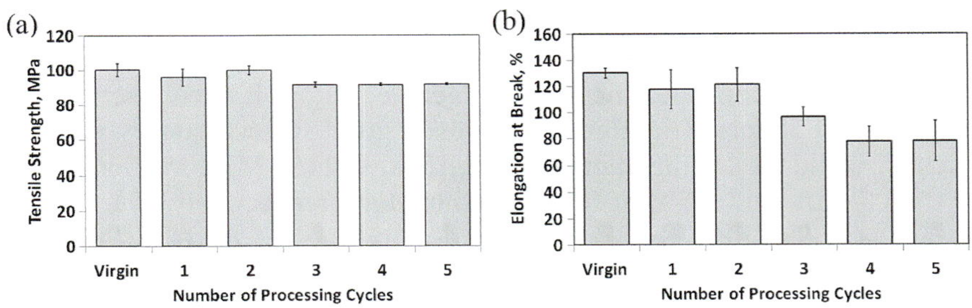

FIGURE 2.9
Tensile strength (a) and elongation at break (b) of regrind PEEK after first, second, third, fourth, and fifth processing cycles.

Source: McLauchlin et al. (2014)

2.6.5 Recycling of PEEK, PEKK, and PEI

The recycling of PAEK or fiber-reinforced PAEK composites is considered important not only for environmental sustainability but also for economic benefits, as they are very expensive materials. In the early study of McLauchlin et al. (2014), the processability of PEEK via injection molding up to five reprocessing cycles of molding/regrinding/remolding was investigated. It was observed that the tensile strength of the first and second regrind PEEK samples did not differ significantly from that of the virgin PEEK, while the tensile strengths of the third, fourth and fifth regrind PEEK samples were about 10% lower (Figure 2.9a). Also, as shown in Figure 2.9b, the elongation at break decreased by 25% after the 3rd cycle and 40% by the 4th cycle, compared to virgin PEEK. In general, the results indicated that recycled PEEK does not suffer significant deterioration in its basic properties up to the first reprocessing cycle and no changes occurred in crystallinity; therefore, it can be 100% reused for injection molding.

In the study of de Seijas et al. (2024), samples of PEEK and PEKK were 3D printed using an FFF device operating under high vacuum. The 3D printed samples were then mechanically recycled and extruded into a filament spool. It was observed that the recycled PEEK filament exhibited an overall good quality, making it suitable for reprocessing. However, at several points, the predefined filament diameter was exceeded. Compared to recycled PEEK, the recycled PEKK filament resulted in higher deviations in filament thickness due to its high melt flow index and the heterogeneity of the feedstock regrind. These observations highlighted the importance of selecting the optimum process parameters – such as heating temperature, extrusion speed, and fan cooling rate – to ensure high quality filament production.

A recycling and remanufacturing technology for 3D printed carbon fiber-reinforced PEEK (CF/PEEK) composites was also presented in the recent work of Zhang (Zhang et al., 2024). Initially, 3D printed samples of virgin CF/PEEK were fabricated using a continuous fiber 3D printer and then were recycled via infrared radiation heating reverse melt recycling. It was observed that the produced 3D printing filament achieved good surface quality with low porosity, while the average tensile strength was 214.99 MPa and 166.60 MPa for the 0.7 mm and 0.8 mm filament diameters, respectively. The new 3D printed samples that were manufactured using the recycled CF/PEEK filament exhibited higher tensile, flexural, impact, and shear strength, while the porosity was reduced from 11.2% to 5.7% compared to the original printed CF/PEEK composites.

A novel scrapping process has also been proposed for recovering panels made from carbon fabric reinforced PEKK and using them in compression molding (Sandá et al., 2023). According to the results, the tensile strength and modulus of elasticity for the recycled samples were slightly lower by 11% and 13%, respectively, compared to the virgin panels. Concerning recyclability, material recovery from the virgin panels into scrap tapes was estimated to 84%, while the fiber mass content of the scrap tapes was similar to that of the virgin panels (approximately 86 wt%).

PEI is an amorphous high-temperature thermoplastic, commonly referred by its trade name ULTEM™. ULTEM 9085 and ULTEM 1010 are two well-known PEI filaments and numerous studies have investigated their durability and quality (Nguyen et al., 2022; Vidakis et al., 2024). ULTEM 1010 exhibits lower printability in terms of interlayer adhesion and porosity, while ULTEM 9085 is a blend of PEI with PC which has been added to improve printability. Recycled PEI is not yet commercially available in the form of 3D printing filament due to the material's high heat resistance and toughness, which can complicate the melting and filament extrusion processes. To the authors' knowledge, no studies have been published so far on the processing of PEI waste into MEX printing filament.

However, composite materials made from PEI and recycled carbon fibers have been investigated and used to manufacture standard samples through injection molding (Hu et al., 2023) and fused filament fabrication after being extruded into filaments (Arslan et al., 2025). The mechanical performance of the fabricated samples was found to be comparable to that of the virgin materials, indicating strong potential for components for related applications.

2.7 Conclusions and Future Research Directions

To ensure product quality and performance while developing in accordance with environmental sustainability, the industry needs to emphasize resource efficiency through the appropriate use of natural resources to minimize waste, and to employ advanced materials and technologies that promote sustainability. Some of the most promising practices which can result in both economic and environmental benefits are the replacement of metals with plastic-based materials, the use of lead manufacturing methods, and the adoption of a circular economy model to extend the life of a product through recycling and remanufacturing.

By integrating recycling technology and material extrusion (MEX) three-dimensional (3D) printing, polymer-based waste can be converted into filament feedstock which will be thereafter employed for the fabrication of functional products. These products made of recycled material can be substituted for their original counterparts without significant loss in performance, and thus retain their value. Furthermore, the development of suitable additives can facilitate the recycling process by enhancing their physical and mechanical properties, as well as improving the recyclability and printability of polymers.

The characterization of 3D printed specimens made from recycled thermoplastic filaments has been investigated in several studies. However, there is a very limited number of studies regarding the recycling of high performance polymers (HPPs) waste into MEX filament suitable for the remanufacturing (3D printing) of components. The main challenges are related to the influence of multiple process parameters on the quality and properties (physical, rheological, mechanical) of the recycled filament and, accordingly, the final 3D printed parts. These parameters are not standard but are user-defined at each stage of the process, such as grinding and melting for mechanical recycling, filament extrusion, and 3D printing. In addition, the process parameters are mostly determined through trial and error, taking into account the filament supplier's recommended values, respective experimental studies in the literature, as well as individual testing, and they do not necessarily ensure the optimal process and performance of the material and the 3D printed part.

To address the slow adoption of circular manufacturing by industry, regarding the recycling and remanufacturing (3D printing) of HPPs, what future research directions and insights should be pursued?

First, there is a need for more studies to provide systematic and validated information and ensure experimental repeatability regarding the recycled thermoplastic filaments and the remanufactured (3D printed) samples. Thermal and rheological properties – such as viscosity, melt flow index, coefficient of thermal expansion, and the temperature and shear dependent flow behavior – must be investigated under varying recycling process parameters. The different behaviors of semi-crystalline HPPs (such as PEEK, PPS, and PVDF) and amorphous ones (such as PEI, PSU, PPSU) needs to be further investigated, as the former are more mechanically stable at high temperatures, while the latter lead to better interlayer bonding.

Moreover, extensive experimental studies with larger sample sizes need to be conducted regarding certain mechanical properties that are critical for industrial environments, such as tensile strength, flexural strength, modulus of elasticity, impact toughness, fatigue, and creep. The interaction between process parameters is quite significant and has an impact on the structural, mechanical, and thermal properties of the final 3D printed parts. The variability in these critical properties must be examined under different conditions in order to understand how HPPs respond to the various parameters at each stage of mechanical recycling and MEX printing process.

To analyze experimental data and examine the process parameters interactions, the design of experiment (DOE) via the Taguchi and Anova methods can been employed. These systematic methods allow for the determination of the relationship between input variables (process parameters) and output variables, minimizing the number of experimental tests required and saving time and costs (Hartig et al., 2022). In the recent work of Refat (Refat et al., 2025), the effect of various parameters of a single screw extruder – such as the barrel temperature, nozzle and cooling bath temperature, screw speed, nozzle diameter, water bath length, and distance to the nozzle – on the properties of a PLA filament and 3D-printed PLA samples was investigated. A Taguchi experimental design was implemented and the optimum parameters for low filament roundness, low surface roughness, and better mechanical properties were obtained.

Furthermore, to address the challenge of large experimental datasets, the employment of artificial intelligence techniques – such as machine learning, deep learning, etc. – can automate additive manufacturing (AM) processes such as material selection, process parameters optimization, and performance prediction. Machine learning models can be trained to develop relationships between input data (i.e., design requirements, process parameters, experimental or simulation results) and critical performance output targets (i.e., physical, mechanical, thermal, surface roughness, dimensional and geometric accuracy properties) and predict the quality and properties of materials and 3D printed samples (Kim et al., 2025; Athanasiou et al., 2024). This improves decision-making regarding new and sustainable materials that are optimized for specific AM applications, while reducing the number of trials, saving time and minimizing material waste.

Also, to ensure the performance of remanufactured parts made from recycled HPPs and the retention of their properties under harsh industrial environmental conditions, additional tests – such as resistance to chemicals, radiation and high thermal loads – should be carried out. Finally, in order to demonstrate the full potential of using recycled filaments as raw materials for remanufacturing functional components, life cycle assessment and cost–benefit analysis should be carried out to verify economic regeneration and efficient environmental performance throughout their life cycle.

Under this perspective, product designers could reconsider the entire manufacturing process and adopt MEX AM in combination with the use of recycled thermoplastics to produce functional end-use products with properties that can be comparable to or even better than their machined or injection-molded counterparts. In addition, significant benefits could arise, such as creating value from end-of-life products, using fewer raw materials, generating less waste, and reducing the environmental impact of products and processes in the manufacturing industry.

References

Al-Mazrouei, N, Al-Marzouqi, AH, Ahmed, W. Characterization and sustainability potential of recycling 3D-printed nylon composite wastes. *Sustainability*. 2022; 14:10458. https://doi.org/10.3390/su141710458

Andrearczyk, A, Konieczny, B, Sokołowski, J. Additively manufactured parts made of a polymer material used for the experimental verification of a component of a high-speed machine with an optimised geometry – preliminary research. *Polymers*. 2021; 13:137. https://doi.org/10.3390/polym13010137

Arslan, D, Mihai, M, Therriault, D, Lévesque, M. Formulation and characterization of polyetherimide composites reinforced with recycled carbon fibers and thermal black particles for fused filament fabrication. *Composites Part A: Applied Science and Manufacturing*. 2025; 194:108946. https://doi.org/10.1016/j.compositesa.2025.108946.

Athanasiou, CE, Liu, X, Gao, H. A perspective on democratizing mechanical testing: harnessing artificial intelligence to advance sustainable material adoption and decentralized manufacturing. *ASME Journal of Applied Mechanics* 2024; 91(11): 110801. https://doi.org/10.1115/1.4066085

Bertagna, S, Braidotti, L, Laurini, E, et al. Thermoplastic materials for the metal replacement of non-structural components in marine engines. *Applied Sciences* 2022; 12:8766. https://doi.org/10.3390/app12178766

Boado-Cuartero, B, Pérez-Álvarez, J, Roibás-Millán, E. Material characterization of high-performance polymers for additive manufacturing (AM) in aerospace mechanical design. *Aerospace.* 2024; 11:748. https://doi.org/10.3390/aerospace11090748

Costa, AA, Martinho, PG, Barreiros, FM. Comparison between the mechanical recycling behaviour of amorphous and semicrystalline polymers: a case study. *Recycling.* 2023; 8:12. https://doi.org/10.3390/recycling8010012

Costa, C, Lopes, P, Castro, J, et al. High-performance thermoplastics obtained by fused filament fabrication: effects of harsh environmental conditions of power transformers. *Progress in Additive Manufacturing.* 2024. https://doi.org/10.1007/s40964-024-00614-z

Cress, AK, Huynh, J, Anderson, EH, et al. Effect of recycling on the mechanical behavior and structure of additively manufactured acrylonitrile butadiene styrene (ABS). *Journal of Cleaner Production.* 2021; 279:123689. https://doi.org/10.1016/j.jclepro.2020.123689

Cruz, F, Lanza, S, Boudaoud, H, et al. *Polymer recycling and additive manufacturing in an open source context: optimization of processes and methods.* Proceedings of the 2015 Annual International Solid Freeform Fabrication Symposium-an Additive Manufacturing Conference, Austin, TX, USA, 2017.

Das, A, Chatham, CA, Fallon, JJ, Zawaski, CE, et al. Current understanding and challenges in high temperature additive manufacturing of engineering thermoplastic polymers. *Additive Manufacturing.* 2020; 34:101218. https://doi.org/10.1016/j.addma.2020.101218

de Leon, ACC, da Silva, IGM, Pangilinan, KD, et al. High performance polymers for oil and gas applications. *Reactive and Functional Polymers.* 2021; 162:104878. https://doi.org/10.1016/j.reactfunctpolym.2021.104878

de Seijas, MOV, Piskacev, M, Celotti, L, et al. Closing the loop in space 3D printing: effect of vacuum, recycling, and UV aging on high performance thermoplastics produced via filament extrusion additive manufacturing. *Acta Astronautica.* 2024; 219:164–176. https://doi.org/10.1016/j.actaastro.2024.03.015

Deshmukh, D, Kulkarni, H, Srivats, DS, et al. Recycling of acrylonitrile butadiene styrene (ABS): a review. *The Polymer Bulletin* 2024; 81:1–38. https://doi.org/10.1007/s00289-024-05269-y

Dimitrellou, S, Strantzali, E, Iakovidis, I. A decision-making strategy for selection of FDM-based additively manufactured thermoplastics for industrial applications based on material attributes. *Sustainable Futures.* 2025; 9:100640. https://doi.org/10.1016/j.sftr.2025.100640

Dobrzańska-Danikiewicz, AD, Siwczyk, B, Bączyk, A, Romankiewicz, A. Mechanical properties of recycled PLA and PETG printed by FDM/FFM method. *Journal of Achievements in Materials and Manufacturing Engineering.* 2023; 119(2):49–59. https://doi.org/10.5604/01.3001.0053.9490

Dollischek, LB, Frohn-Sorensen, P, Kurz, F, Engel, B. Injection molding of post-industrial recycled glass fiber reinforced polyphenylene sulfide (PPS GF40): industrial feasibility and material analysis. *Journal of Cleaner Production.* 2024; 471:143416. https://doi.org/10.1016/j.jclepro.2024.143416

Garmabi, MM, Shahi, P, Tjong, J, Sain, M. 3D printing of polyphenylene sulfide for functional lightweight automotive component manufacturing through enhancing interlayer bonding. *Additive Manufacturing.* 2022; 56:102780. https://doi.org/10.1016/j.addma.2022.102780.

Geslin, A, Paul, C, Bussi, P. PEKK poly ether ketone ketone for high temperature high pressure oil & gas conditions. *Offshore Technology Conference.* 2021; OTC-30987-MS. https://doi.org/10.4043/30987-MS

Gültürk, C, Berber, H. Effects of mechanical recycling on the properties of glass fiber–reinforced polyamide 66 composites in automotive components. *e-Polymers.* 2023; 23:20230129. https://doi.org/10.1515/epoly-2023-0129

Gurchinoff, S, Fish, D, Stern, B. High performance polymers for small engine applications. *JSAE/SAE 2013 Small Engine Technology Conference.* 2013. https://doi.org/10.4271/2013-32-9012

Hartig, S, Hildebrandt, L, Fette, M, et al. Process parameter determination for small recycling plants for the production of filament for FFF printing using the Taguchi method. *Progress in Additive Manufacturing.* 2022; 7:87–97. https://doi.org/10.1007/s40964-021-00218-x

Hasan, MR, Davies, IJ, Paramanik, A, et al. Fabrication and characterisation of sustainable 3d-printed parts using post-consumer PLA plastic and virgin PLA blends. *Processes.* 2024a; 12:760. https://doi.org/10.3390/pr12040760

Hasan, MR, Davies, IJ, Pramanik, A, et al. Impact of process parameters on improving the performance of 3D printed recycled polylactic acid (rPLA) components. *The International Journal of Advanced Manufacturing Technology* 2024b; 131:3751–3779. https://doi.org/10.1007/s00170-024-13140-7

Hu, J-M, Zhou, Z-M, Guo, F-L, et al. Investigation of recycling effects on the mechanical properties of short carbon and glass fiber reinforced polyetherimide composites. *Journal of Applied Polymer Science.* 2023; 140(21). https://doi.org/10.1002/app.53871

Husain, M, Singh, R, Pabla, B. On Secondary Recycling of PVDF for 3D Printing of Functional Sensors: Mechanical, Flow, Thermal, and Dielectric Properties. *National Academy Science Letters.* 2023; 47(11). https://doi.org/10.1007/s40009-023-01364-2

Karthick, N, Soundararajan, R, Arul, R, Prasanth, JR. Evolution of tribological performance of polypropylene with carbon fibre composites fabricated through FDM technology by varying infill density. *The Journal of the Institution of Engineers (India): Series D.* 2024; 105(2):961–968. https://doi.org/10.1007/s40033-023-00503-9

Kim, SH, Park, JH, Park, JY, et al. Development of machine learning models for material classification and prediction of mechanical properties of FDM 3D printing outputs. *Journal of Mechanical Science and Technology* 2025; 39:541–552. https://doi.org/10.1007/s12206-024-1114-9

Kumar, E, Ozcelik, B, Yilmaz, F. The effects of recycling process on thermal, chemical, rheological, and mechanical properties of PC/ABS binary and PA6/PC/ABS ternary blends. *Journal of Elastomers & Plastics.* 2015; 48(2). https://doi.org/10.1177/0095244315576239

Kurdi, A, Chang, L. Recent advances in high performance polymers – tribological aspects. *Lubricants.* 2019; 7(2). https://doi.org/10.3390/lubricants7010002

Li, Z, Yu, L, Yan, Z. Effect of hydrothermal aging on mechanical properties and long-term durability of recycled Polysulfone. *Journal of Polymers and the Environment.* 2022; 30:92–101. https://doi.org/10.1007/s10924-021-02182-7

Lodha, S, Song, B, Park, SI, et al. Sustainable 3D printing with recycled materials: a review. *Journal of Mechanical Science and Technology.* 2023; 37:5481–5507. https://doi.org/10.1007/s12206-023-1001-9

Maletzko, C, Herz, N. *Polyarylsulfones (PSU, PESU, PPSU)*. Kunststoffe International, Carl Hanser Verlag, 2011. Available in https://download.basf.com/p1/8a8081c57fd4b609017fd65cb50f6e07/en.

McLauchlin, AR, Ghita, OR, Savage, L. Studies on the reprocessability of poly(ether ether ketone) (PEEK). *Journal of Materials Processing Technology*. 2014; 214:75–80. http://dx.doi.org/10.1016/j.jmatprotec.2013.07.010

Mehta, A, Singh, P, Pabla, BS, Kumar, V. On 3D printed polyvinylidene fluoride-based smart energy storage devices. *Journal of Thermoplastic Composite Materials*. 2024; 37(6):1921–1937. https://doi.org/10.1177/08927057231208133

Mishra, PK, Jagadesh, T. Applications and challenges of 3D printed polymer composites in the emerging domain of automotive and aerospace: a converged review. *The Journal of The Institution of Engineers (India): Series D*. 2022. https://doi.org/10.1007/s40033-022-00426-x

Mishra, V, Negi, S, Kar, S. FDM-based additive manufacturing of recycled thermoplastics and associated composites. *Journal of Material Cycles and Waste Management*. 2023a; 25:758–784. https://doi.org/10.1007/s10163-022-01588-2

Mishra, V, Ror, CHK, Negi, S, Kar, S, Borah, LN. 3D printing with recycled ABS resin: effect of blending and printing temperature. *Materials Chemistry and Physics*. 2023b; 309:128317. https://doi.org/10.1016/j.matchemphys.2023.128317

Montagna, LS, Kondo, MY, Callisaya, ES, et al. A review on research, application, processing, and recycling of PPS based materials. *Polímeros: Ciência e Tecnologia*. 2022; 32(1): e2022005. https://doi.org/10.1590/0104-1428.20210108

Morales, MA, Maranon, A, Hernandez, C, et al. Colombian sustainability perspective on fused deposition modeling technology: opportunity to develop recycled and biobased 3D printing filaments. *Polymers*. 2023; 15:528. https://doi.org/10.3390/polym15030528

Morseletto, P. Targets for a circular economy. *Resources, Conservation & Recycling*. 2020; 153:104553. https://doi.org/10.1016/j.resconrec.2019.104553

Nagengast, N, Bay, C, Döppe, RF, et al. Thermo-mechanical recyclability of additively manufactured polypropylene and polylactic acid parts and polypropylene support structures. *Polymers*. 2023; 15:2291. https://doi.org/10.3390/polym15102291

Nguyen, KQ, Vuillaume, PY, Robert, M, Elkoun, S. *AFM analysis of 3D printing PEI for automotive applications*. Proceedings of the 8th International Conference on Mechanical, Automotive and Materials Engineering, 2022. https://doi.org/10.1007/978-981-99-3672-4_10

Olscher, C, Jandric, A, Zafiu, C, Part, F. Evaluation of marker materials and spectroscopic methods for tracer-based sorting of plastic wastes. *Polymers*. 2022; 14. https://doi.org/10.3390/polym14153074

Pervaiz, S, Qureshi, TA, Kashwani, G, Kannan, S. 3D printing of fiber-reinforced plastic composites using fused deposition modeling: a status review. *Materials*. 2021; 14:4520. https://doi.org/10.3390/ma14164520

Pickering, K, Stoof, D. Sustainable composite fused deposition modelling filament using post-consumer recycled polypropylene. *Journal of Composites Science*. 2017; 1(2): 17. https://doi.org/10.3390/jcs1020017

Prado, AR, Leal-Junior, AG, Marques, C, et al. Polymethyl methacrylate (PMMA) recycling for the production of optical fiber sensor systems. *Optics Express*. 2017; 25(24):30051–30060. https://doi.org/10.1364/OE.25.030051

Punia, R, Kandasubramanian, B. Towards circular economy: integrating polymer recycling and 3d printing for economic regeneration. *Polytechnica*. 2025; 8:2. https://doi.org/10.1007/s41050-025-00054-w

Rashid, AA, Koç, M. Additive manufacturing for sustainability and circular economy: needs, challenges, and opportunities for 3D printing of recycled polymeric waste. *Materials Today Sustainability*. 2023; 24:100529. https://doi.org/10.1016/j.mtsust.2023.100529

Razavykia, A, Delprete, C, Rosso C, Baldissera, P. Functionality analysis of thermoplastic composite material to design engine components. *SAE Technical Paper*. 2020. https://doi.org/10.4271/2020-01-0774.

Refat, M, Maertens, R, Weiss, P, et al. Investigation of the influence of manufacturing on filament production and its impact on additive manufactured structures. *Polymers*. 2025; 17(5):651. https://doi.org/10.3390/polym17050651

Reich, MJ, Woern, AL, Tanikella, NG, Pearce, JM. Mechanical properties and applications of recycled polycarbonate particle material extrusion-based additive manufacturing. *Materials*. 2019; 12(10):1642. https://doi.org/10.3390/ma12101642

Romani, A, Perusin, L, Ciurnelli, M, Levi, M. Characterization of PLA feedstock after multiple recycling processes for large-format material extrusion additive manufacturing. *Materials Today Sustainability*. 2024; 25:100636. https://doi.org/10.1016/j.mtsust.2023.100636

Sam-Daliri, O, Ghabezi, P, Steinbach, J, et al. Experimental study on mechanical properties of material extrusion additive manufactured parts from recycled glass fibre-reinforced polypropylene composite. *Composites Science and Technology*. 2023; 241:110125. https://doi.org/10.1016/j.compscitech.2023.110125

Sandá, A, Ruiz, R, Mafé, MA, et al. Scrapping of PEKK-based thermoplastic composites retaining long fibers and their use for compression molded recycled parts. *AIMS Materials Science*. 2023; 10(5):819–834. https://doi.org/10.3934/matersci.2023044.

Shanmugam, V, Das, O, Neisiany, RE, et al. Polymer recycling in additive manufacturing: an opportunity for the circular economy. *Materials Circular Economy*. 2020; 2:11. https://doi.org/10.1007/s42824-020-00012-0

Singh, R, Kumar, R, Ranjan, N. Sustainability of recycled ABS and PA6 by banana fiber reinforcement: thermal, mechanical and morphological properties. *Journal of The Institution of Engineers (India): Series C* 2019; 100(2):351–360. https://doi.org/10.1007/s40032-017-0435-1

Sponchioni, M, Altinok, S. Poly(methyl methacrylate): market trends and recycling. In: Moscatelli D, Pelucchi M, editors. *Advances in Chemical Engineering*. Academic Press; 2022;60(1):269–287.

Strano, M, Farioli, D, Di Landro, LA, Mussi, V. Printability of recycled feedstock for extrusion additive manufacturing. *Advances in Materials and Processing Technologies*. 2023. https://doi.org/10.1080/2374068X.2023.2295651

Su, N, Zhu, C, Chen, X. Investigation of tensile properties of FDM-printed reclaimed carbon fiber reinforced reclaimed polyamide-6 composites (rCF/rPA) via various densities of infill patterns. *Journal of Physics: Conference Series*. 2024; 2679:012033. https://doi.org/10.1088/1742-6596/2679/1/012033

Tanney, D, Meisel, NA, Moore, J. *Investigating material degradation through the recycling of PLA in additively manufactured parts*. Proceedings of the 28th Annual International Solid Freeform Fabrication Symposium – an Additive Manufacturing Conference, 2017, 519–531.

Tinz, J, de Ancos, T, Rohn, H. Carbon footprint of mechanical recycling of post-industrial plastic waste: study of ABS, PA66GF30, PC and POM regrinds. *Waste*. 2022; 1:127–139. https://doi.org/10.3390/waste1010010

Türker, YS, Öztürk, F, Öz, Y. Review of recycling methods of thermoplastic composite materials. *Polymer-Plastics Technology and Materials*. 2024; 63(12):1693–1713. https://doi.org/10.1080/25740881.2024.2352148

Vidakis, N, Petousis, M, Michailidis, N, Papadakis, V, et al. Cyclic economy driven additive manufacturing: valorization of mechanically recycled poly(methyl methacrylate) scrap in material extrusion 3D printing. *Journal of Cleaner Production*. 2025; 486:144639. https://doi.org/10.1016/j.jclepro.2024.144639.

Vidakis, N, Petousis, M, Mountakis, N, et al. Energy consumption vs. tensile strength of poly[methyl methacrylate] in material extrusion 3D printing: the impact of six control settings. *Polymers*. 2023; 15:845. https://doi.org/10.3390/polym15040845

Vidakis, N, Petousis, M, Spiridaki, M, et al. Optimization of critical process control parameters in MEX additive manufacturing of high-performance polyethylenimine: energy expenditure, mechanical expectations, and productivity aspects. *The International Journal of Advanced Manufacturing Technology* 2024; 132:1163–1192. https://doi.org/10.1007/s00170-024-13418-w

Vidakis, N, Petousis, M, Tzounis, L, et al. Sustainable additive manufacturing: mechanical response of polyamide 12 over multiple recycling processes. *Materials*. 2021a; 14:466. https://doi.org/10.3390/ma14020466

Vidakis, N, Petousis, M, Tzounis, L, et al. Sustainable additive manufacturing: mechanical response of polyethylene terephthalate glycol over multiple recycling processes. *Materials*. 2021b; 14:1162. https://doi.org/10.3390/ma14051162

Vieyra, H, Molina-Romero, JM, Calderón-Nájera, JDD, Santana-Díaz, A. Engineering, recyclable, and biodegradable plastics in the automotive industry: a review. *Polymers*. 2022; 14:3412. https://doi.org/10.3390/polym14163412

Weyhrich, CW, Long, TE. Additive manufacturing of high-performance engineering polymers: present and future. *Polymer International*. 2022; 71:532–536. https://doi.org/10.1002/pi.6343

Yang, C-J, Shih, H-C, de Ancos, T. Comparison of human health, energy consumption, and carbon emission with virgin and recycled filament in fused deposition modeling printing. *Materials Today Sustainability*. 2024; 25:100640. https://doi.org/10.1016/j.mtsust.2023.100640

Yu, L, Zhao, D, Wang, W. Mechanical properties and long-term durability of recycled polysulfone plastic. *Waste Management*. 2019; 84:402–412. https://doi.org/10.1016/j.wasman.2018.11.025

Zhan, X, Su, K, Tuo, X, Gong, Y. Mechanical, thermal, and electrical properties on 3D printed short carbon fiber reinforced polypropylene composites. *ACS Applied Polymer Materials Current Issue*. 2024; 6:3787–3795. https://doi.org/10.1021/acsapm.3c03083

Zhang, M, Tian, X, Jia, L, et al. Mechanism and performance of thermal radiation recycling for in-space 3D printed continuous carbon fibre-reinforced PEEK composites. *Virtual and Physical Prototyping*. 2024; 19(1):e2425823. https://doi.org/10.1080/17452759.2024.2425823

Zhang, M, Tian, X, Liu, T, et al. Thermal degradation and performance evolution mechanism of fully recyclable 3D printed continuous fiber self-reinforced composites. *Sustainable Materials and Technologies*. 2023; 37:e00706. https://doi.org/10.1016/j.susmat.2023.e00706

Zhang, Y, Kristinsson, VT, Uríček, M. Effects of process parameters on print quality and mechanical properties of FDM printed polyoxymethylene samples. *21st International Conference of the European Society for Precision Engineering and Nanotechnology*. 2021; Copenhagen.

3

Integrating Circular Economy Principles in Ship Recycling: A Review of Decision-Making Approaches

Eleni Strantzali, Epaminondas D. Banos, Sotiria Dimitrellou, Georgios A. Livanos, and Dimitrios-Nikolaos S. Pagonis

3.1 Introduction

Maritime transport constitutes a crucial part of the global economy, with over 80% share of the volume of goods carried by sea (Strantzali et al., 2023). However, environmental and social challenges are revealed when a ship reaches the end of its operational life as it faces scrapping, recycling, or even abandonment. Shipbreaking has traditionally been carried out in a way that compromises workforce safety and environmental impact. Currently, the main recycling facilities are in South Asia, where countries are attractive to ship owners due to low labor costs and loose environmental regulations. Instead, recycling contains recyclable materials, like copper, steel, electronics, etc., which can contribute to the broader efforts in sustainable waste management (ElMenshawy et al., 2024; Moussa et al., 2024).

Recycling falls under the main principles of the circular economy (CE) approach (Tola et al., 2023). The adoption of circular economy principles has been expanding since the late 1970s. Having studied the definitions of CE in the literature, Geissdoerfer et al. (2017) proposed the following definition:

> Circular Economy is a regenerative system in which resource input and waste, emission, and energy leakage are minimized by slowing, closing, and narrowing material and energy loops. This can be achieved through long-lasting design, maintenance, repair, reuse, remanufacturing, refurbishing, and recycling.

Focusing on ships, it is estimated that 95–98% of their light displacement tonnage (LTD) are recyclable materials (Tola et al., 2023).

DOI: 10.1201/9781003654889-3

During the last decade, the field of ship recycling has attracted the interest of many researchers in the academic community. Shipowners, regulatory bodies, recycling yard operators, and environmental groups have diverse priorities when it comes to facing the problem of ship dismantling. To address these conflicting preferences, decision-making methodologies have been adopted. They offer the flexibility and capacity required to deal with complex and multi-objective issues, providing robust, optimal options. As a result, they facilitate more balanced and transparent decisions that align with long-term sustainability goals in the ship recycling industry (Moussa et al., 2024; Mannan et al., 2022).

The goal of this chapter is to investigate this field and demonstrate the current state of the literature. It seeks to identify trends and challenges, while also exploring the effectiveness of various decision-making methodologies in addressing the complexities of ship recycling. This review analysis aims to provide insights for future research and policy development.

This chapter is organized as follows: Section 3.2 analyzes ship recycling and outlines the applicable legislation. Section 3.3 describes the link between the CE and ship recycling, exploring how these two concepts are interconnected and the implications they have for the maritime industry. The implementation of decision-making methodologies is the focus of Section 3.4, which is then followed by the conclusions of Section 3.5.

3.2 Ship Recycling

The ship recycling process is described as the technical operation of dismantling and recycling a vessel at the end of its operational life. The term mainly refers to the activity of breaking down a vessel into smaller sections and pieces that finally are delivered to the shoreside and the raised materials to be disposed mainly into the CE and serve various purposes (Jain et al., 2013).

The ship recycling process can be characterized as an environmentally sound and sustainable development operation, as it supports the reduction of the need of natural resources. Tons of marine-type steel recovered during the breaking of a ship, added with the marine equipment secondary market, represents 98% of the ship weight, making the activity well worthwhile for the disposal of a ship at its end of operational life (Jain et al., 2013).

In 2001, the Organisation for Economic Co-operation and Development (OECD) noted an increasing casualty rate for older ships remaining in operation, calling for the steady withdrawal of older ships and their replacement by new tonnage characterized by environmentally friendly designs, greater operating efficiency, and a general reduction in marine risk (IMO Resolution A.962[23]).

The maritime market, economic, regulatory, and environmental governing factors, as well global geopolitical crisis and uncertainty (Tuttle, 2024), drive

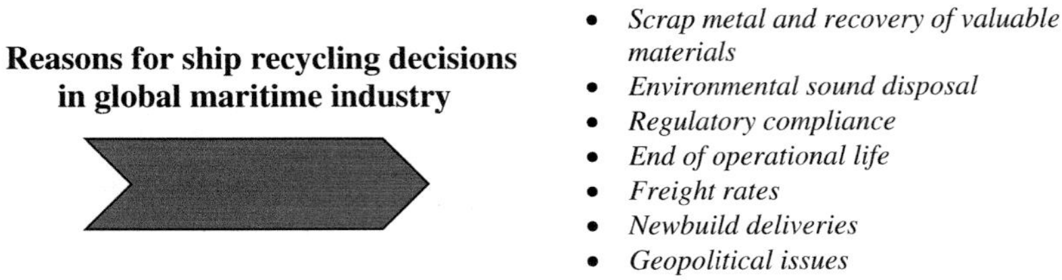

FIGURE 3.1
Reasons for ship recycling decisions.

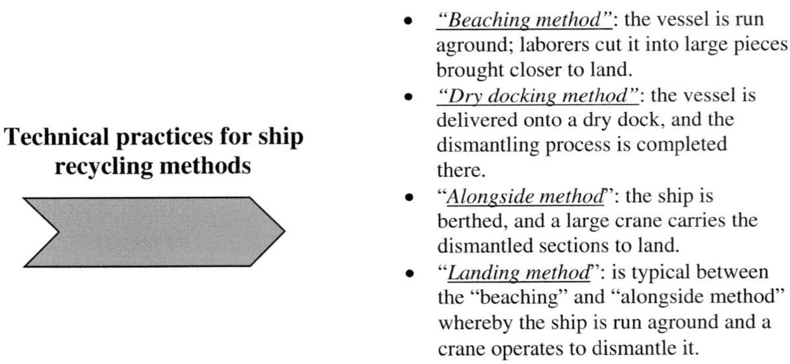

FIGURE 3.2
Ship recycling methods.

the decision of a ship owner to recycle its fleet independently if it is at the end of the operation life. As noted in what follows, the activity of the ship recycling sector is crucial for the global market of the shipping industry, and the decision by the stakeholders to recycle or not is driven by different incentives (Figure 3.1).

Referring to technical common practices for ship recycling, four main techniques have been developed that set the provisions of dismantling the whole vessel into smaller sections. Each technique has different characteristics and is applied under local circumstances such as environmental regulatory frameworks, typical local tide profiles, and shipyard installed infrastructure. Each of these ship recycling techniques (Figure 3.2) has different strong and weak characteristics concerning the environment and labor laws.

In terms of working conditions, ship recycling is a complex industry with significant implications in the view of safety. However, increasing awareness, regulations, and initiatives are pushing for improvements. The International Labor Organization of the United Nations (UN) has developed a framework on safety and health in shipbreaking to assist competent authorities and shipbreaking facilities in implementing relevant safety and health standards (Sarraf et al., 2010).

Consequently, ship recycling is driven by the need to recover valuable resources, manage environmental risks associated with obsolete vessels, comply with regulations, and provide a responsible end-of-life solution for ships that are no longer economically viable. While the industry has faced criticism for unsafe and environmentally damaging practices in some regions, increasing regulation and a growing focus on sustainability are pushing for safer and cleaner ship recycling methods.

The maritime sector clarifies that the quantities of vessels to be recycled per year can fluctuate based on uncertain factors within the shipping industry. Historically, around 400 large ships were estimated to be dismantled annually worldwide, and in 2024, the total gross tonnage (GT) of recycled merchant ships was around 4.6 million GT (Lloyd's List, 2025).

The NGO Shipbreaking Platform organization states that Bangladesh, India, and Pakistan delivered the majority of vessels for recycling purpose during 2024, estimated around 80%, (NGO Shipbreaking Platform, website); according to studies by the Baltic and International Maritime Council (BIMCO), it is estimated that more than 15,000 ships could be recycled in the next decade (BIMCO, website). Finally, BCC Research records that the global ship recycling sector projections estimates the market to reach $11.5 billion by 2028 (BCC Research, website).

According to the latest records, the first semester of 2025, almost 78 large ships (tankers, dry goods, container and gas carriers) were sold for scrapping at the Asian yards. That number of vessels remains the lowest of the last years which reflects the high risk uncertainty of the global market to set new buildings and retrieve the older vessels (Tsatsaronis, 2025).

3.2.1 Current Regulatory Framework

Ship recycling operation is characterized as one of the most hazardous processes in terms of human occupations and environmental issues. A robust regulatory framework has been developed to ensure that the process is framed by modules capable to minimize local adverse environmental impacts and to ensure a low occupational health and safety risks related to ship recycling industry.

3.2.1.1 IMO, Hong Kong International Convention (HKC, 2009)

Ship scrapping and recycling issues were first raised at the 44th MEPC (Marine Environment Protection Committee) session of IMO, in March 2000. Guidelines were developed and adopted as *Guidelines on Ship Recycling* by Resolution A.962(23) that supports all the maritime industry including flag administrations of ship building, as well as commercial bodies such as shipowners, ship builders, repairers, recycling yards, and manufactories. In 2009, the final Act of the Hong Kong International Convention (HKC) for the Safe and Environmentally Sound Recycling was adopted by IMO Conference.

That Convention is entering into force on 26 June 2025 and is surrounded by IMO's resolutions for vessels and shipyards facilities.

HKC (2009) intends to address a life cycle approach of a ship from the conceptual design project of the ship, through to the construction method and materials used. The Convention covers mainly all the issues around ship recycling, including the fact that ships sold for scrap are waste and, in many cases, could be declared as hazardous waste, as they may contain environmentally hazardous substances. That Convention, in its provisions, also concerns labor safety conditions at authorized recycling locations worldwide. Under HKC (2009) provisions, different stakeholders, such as ships and flag states, recycling facilities, recycling states, shipbuilding states, and port states, have certain roles and responsibilities according to the requirements stated under the 21 Articles of the Convention and the 25 regulations of its appendices.

Pursuant to Regulation 5.1 and 5.3 of HKC, each ship flying the flag or calls at a port of an HKC party should sail under the requirement to carry onboard and maintain an inventory of hazardous materials (IHM). The same document should be onboard, even if the ship is slated to be recycled at an authorized ship recycling facility.

Before recycling, the shipowner should provide all the required information to the ship recycling facility to prepare the ship recycling plan which is unique for each vessel. That plan, accompanied by reporting documentation, is communicated to the relevant authorities of the recycling facility to approve or deny the ship recycling plan. Following acceptance of the plan, a final onboard survey is executed by the vessel's flag state authority and an international ready for recycling certificate is issued before the start of recycling on site. After the completion of the recycling process, a statement of recycling completion is endorsed from the ship recycling facility and circulated to the ship recycling country's authorities and the flag state of the vessel flying the flag.

It is stipulated that, beside the others, a structural requirement of that Convention is the existence of: (a) an inventory of hazardous materials certificate onboard; and (b) an international ready for recycling certificate (IRRC), in case that the vessel sails to an facility approved under the Hong Kong Convention.

IMO's MEPC, in the scope of early implementation and enforcement of the Convention's technical standards, has developed and adopted the following guidelines to assist its member States.

- 2003: IMO Guidelines on Ship Recycling, IMO Resolution A.962(23).
- 2011: Guidelines for the Development of the Inventory of Hazardous Materials, IMO Resolution MEPC.197(62).
- 2009: Calculation of Recycling Capacity for Meeting the Entry-into-Force Conditions of the Hong Kong International Convention for the

Safe and Environmentally Sound Recycling of Ships, IMO Resolution MEPC.178(59).

- 2011: Guidelines for the Development of the Ship Recycling Plan, adopted by resolution MEPC.196(62).
- 2012: Guidelines for Safe and Environmentally Sound Ship Recycling, adopted by resolution MEPC.210(63).
- 2012: Guidelines for the Authorization of Ship Recycling Facilities, adopted by resolution MEPC.211(63).
- 2012: Guidelines for the Survey and Certification of Ships under the Hong Kong Convention, adopted by resolution MEPC.222(64).
- 2012: Guidelines for the Inspection of Ships under the Hong Kong Convention, adopted by resolution MEPC.223(64).
- 2023: Guidelines for the Development of the Inventory of Hazardous Materials, adopted by resolution MEPC.379(80).
- 2025: Amendments to the 2023 Guidelines for the Development of the Inventory of Hazardous Materials [Resolution MEPC.379(80)], Resolution MEPC.405(83).

3.2.1.2 UN, the Basel Convention, and "the Ban Amendment"

The Basel Convention (BC), with its 29 structural articles and annexes, constitutes the control of transboundary movements of hazardous wastes and their disposal, especially between developed and less developed countries, in a scope to manage the hazardous wastes in an environment friendly way. Transboundary movements of ships can be associated with the field of application of that Convention in the case that the vessel is characterized as hazardous waste, since it contains hazardous materials (Basel Convention, UN, Revised 2023).

In 1995, the Amendment to the Basel Convention ("the Ban Amendment") was adopted and entered into force on 5 December 2019. The Ban Amendment stipulates the provisions for the prohibition of all transboundary movements of hazardous wastes for disposal, reuse, or recycling by each party included in the proposed new Annex VII to the Convention (parties and other states which are members of the OECD, EU [European Union], Liechtenstein) to states not included in Annex VII (Basel, website). It is important to notice that under IMO and UN cooperation, countries which are parties on both HKC 2009 and Basel Convention should notify the Secretariat of the BC that their ships should not fall under the provisions of BC and "The Ban Amendment," only in the case that they sail to their final recycling destination pursuant to HKC regulations (HKSRC.2/Circ.1, 2024).

3.2.1.3 EU, Ship Recycling Regulation (EU SRR)

- EU Ship recycling Regulation 1257/2013.
- EU Regulation 2024/1157 on shipments of waste.

Similar to the IMO International Hong Kong Convention is the EU Ship Recycling Regulation 1257/2013 (Regulation (EU) 1257/2013) that manages in an environmentally sound manner ship recycling and the storage of hazardous materials. That Regulation is applicable on all ships greater than 500 GT under ITC 1969 flying the flag of an EU member state, and on ship recycling facilities that meet the requirements of the Regulation (Decision [EU] 2025/322). These shipyard facilities are located in both EU and non-EU countries and must comply with a series of requirements related to workers' safety and environmental protections.

Under the rules of the EU Ship Recycling Regulation, all the ships flying an EU member state flag or vessels flying non-EU flags should have onboard a maintained inventory of hazardous materials that exist onboard or wastes which should be disposed of. The flag state in EU compromises the final survey onboard, on the scope of the Regulation, and endorses a ready for recycling certificate which is valid for three months until the final dismantling of the vessel at an authorized facility. A strong notification folder is circulated among the stakeholders such as the ship owner, the EU flag state, the ship recycling shipyard, the European commission, and the ship recycling country until the final issuance of the statement of completion and the report on incidents and accidents that are submitted by the recycling company to the flag state of the vessel.

Additionally to the previously mentioned regulations, the EU Waste Shipment Regulation (WSR), EU Regulation 2024/1157 (Regulation [EU] 2024/1157) on shipments of waste, is a set of rules governing the movement of waste within, to, and from the EU. The Regulation aims to protect the environment and human health by ensuring that all the waste is managed properly. The latest version of the Regulation (EU) 2024/1157 entered into force on 20 May 2024.

Shipowners are required to demonstrate compliance with the WSR requirements and to fully document the recycling decision and process. Any circumvention of the requirements, through reflagging or the use of non-approved facilities, may result in sanctions. Pursuant to its principal articles that the Regulation creates the conditions for more transparent, fair, and sustainable ship recycling in Europe, limiting the export of environmental and social impacts to third countries and strengthening the implementation of CE principles. The main benefits of the EU WSR could be gathered as presented in Figure 3.3.

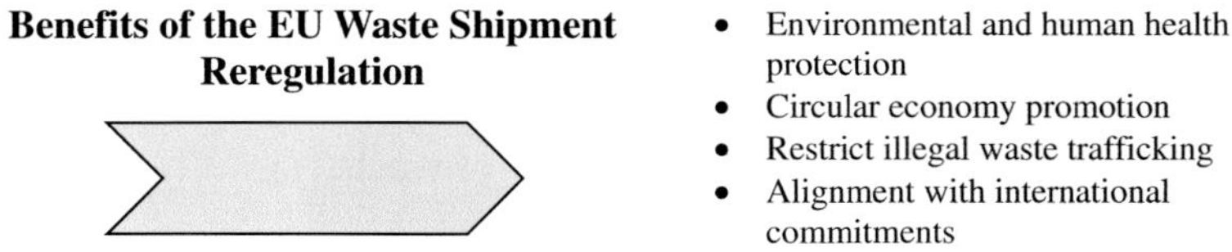

FIGURE 3.3
Benefits of the EU Waste Shipment Reregulation.

For the purposes of the previously mentioned ship recycling process, the European Maritime Safety Agency (EMSA) has circulated to its members technical guidelines published by the following texts.

- EMSA, 2008. Study on the certification of ship recycling facilities.
- EMSA, 2016. Best Practice Guidance on the Inventory of Hazardous Materials.
- EMSA, 2017. Guidance on the Inventory of Hazardous Materials.
- EMSA, 2019. Guidance inspections ship recycling.

For the maritime shipping sector, the benefit of the adoption of these guidelines is that they make it easier to implement the HKC's technical Regulations.

Therefore, the EU WSR aims to create a more sustainable and transparent system for managing waste shipments, prioritizing environmental and health protection and encourage facilitating transition to a CE model within the European Union, (European Commission Waste Shipments, website).

3.3 Circular Economy and Ship Recycling

The CE aims to improve resource utilization and increase value retention, as well as reducing extraction through reuse, remanufacture, and recycling (Tukker, 2015). This is achieved by minimizing raw materials, labor, energy, and capital, and by minimizing environmental impacts, such as carbon dioxide (CO_2) emissions, during manufacturing processes. CE implementation is dominated by the "6R" principles of reduction, reuse, recycling, recovery, redesign, and reconstruction. Although the 6R principles have become more widespread in other transport sectors, the shipping sector presents a mixed pattern, with ship recycling being the dominant practice. Figure 3.4 illustrates the contribution of stakeholders in maritime industry within the CE principles (Okumus et al., 2023; Milios et al., 2019).

CE principles can be leveraged for ship recycling, which is important as it helps to ensure that materials are used responsibly and that the environment

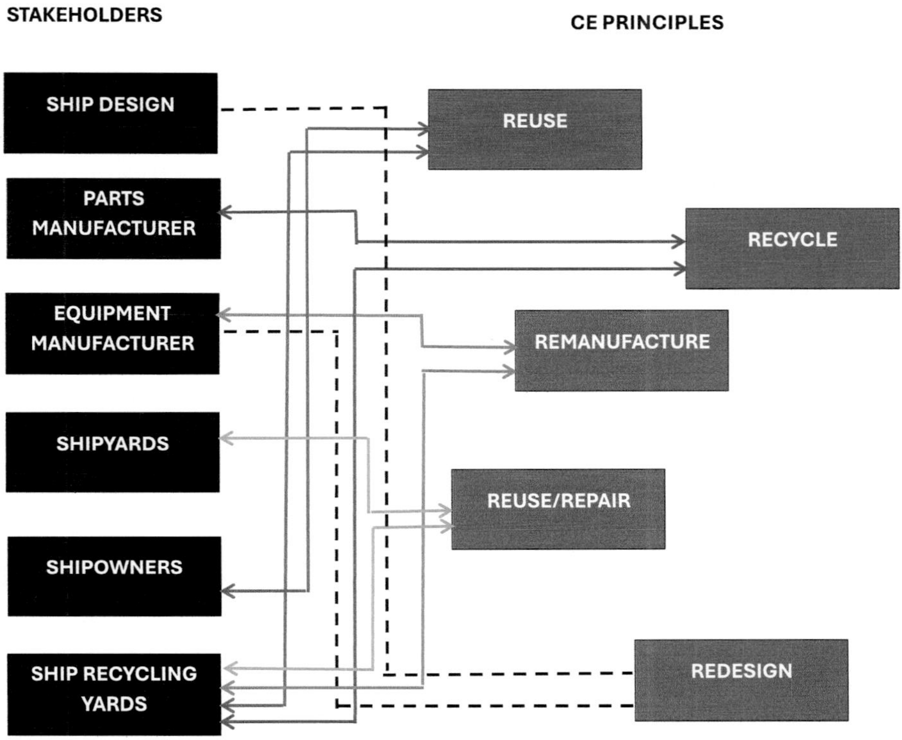

FIGURE 3.4
The contribution of maritime stakeholders in CE principles.

is protected. Approximately 85–90% of the materials in ships can be recoverable, enhancing the role of the ship recycling industry in CE. The transition from linear economy to CE offers enormous potential for sustainable and responsible practices. The ship dismantling industry can achieve more sustainable and efficient operation by adopting approaches such as designing durability and recyclability, maximizing recovery of materials, and developing robust recycling infrastructure. This transition process requires collective effort and collaboration among stakeholders and investments in research and innovation.

Recycled materials can be a driving force in the national economy, as they can be used in the industrial and construction sectors. Ship prices play a vital role in the shipowners' decision for ship recycling. A higher price for an EoL ship, offered by a ship recycling yard, is always attractive. In addition, it is combined with a ship's obsolescence, current earnings, and future market expectations, and determines the supply and demand in ship recycling and freight market dynamics. Ship obsolescence is divided into three cases: physical, technical, and regulatory. Physical obsolescence occurs due to the natural aging process, technical obsolescence due to increased competitiveness by newer and more efficient ships, and regulatory obsolescence caused by the constant upgrading of regulatory requirements. Passenger and container ships have variable dismantling ages, due to rapid technological issues, whereas bulk carriers and tankers have higher dismantling ages because of higher investment cost and operational durability (ElMenshawy et al., 2024).

3.3.1 Recyclable Materials

Ship recycling is a complex and regulated concept, which involves dismantling the vessel and managing a variety of materials and components through recycling, reuse, or disposal, contributing to waste reduction, extending the life cycle of materials, and preventing environmental pollution from hazardous materials.

Common valuable elements can be recovered include ferrous and non-ferrous metals, machinery components, electrical and electronic equipment, appliances and devices, furnishing, insulation materials, etc. Steel from the ship's hull can be melted down or rerolled and reused in construction, while non-ferrous metals such as aluminum, copper alloys, titanium alloys, and bronze can be used instead of raw materials and recovered for their value.

In addition, components and parts from removed products can be repaired or refurbished and used to remanufacture new products with the same function or be repurposed for new products with a different function. Reusable items, such as marine spare parts, communication devices, accommodation and office equipment, etc., can be resold to various small-scale enterprises, refurbishment shops, second-hand markets, and spare parts suppliers (ElMenshawy et al., 2024; Mannan et al., 2024).

According to Tola et al. (2023), it is estimated that recyclable materials of an end-of-life ship represents 95–98% of its light displacement tonnage (LTD). Most materials and components extracted from ships can be reused either in shipbuilding or in various applications in other industries, minimizing the extraction of natural resources and reducing the energy consumption and pollutant emissions generated by the extraction of raw material and the manufacture of new products.

The most important material recycled and reused in construction, shipbuilding, and other manufacturing processes is steel scrap. According to Mannan et al. (2024), using steel scrap instead of iron ore results in a reduction in greenhouse gas (GHG) emissions, as energy consumption in steel manufacturing is reduced by up to 70%.

Steel scrap obtained from the dismantling of a ship consists of irregular pieces of steel which are further melted down in furnaces and used to produce steel products in foundries, and rerollable structural components such as plates, girders, beams, angle bars, etc. The latter are much more valuable than melted scrap, as rerolled products are widely used in the construction sector and especially for the construction of new ships (Mannan et al., 2024). According to interviews with corporate stakeholders in China (Steuer et al., 2021), recovered steel is of high quality and approximately 20% can be used directly in construction projects, while the remaining 80% can be used in the construction of new ships after rerolling.

Similarly, steel, stainless steel, and other valuable non-ferrous metals – such as aluminum, copper, bronze, lead, titanium, copper/nickel, and their alloys – can be recovered from hatchway covers, tanks, valves, pipes, and other equipment and devices, and processed for recycling.

Findings from ship recycling practices in various regions (Steuer et al., 2021) showed that steel constitutes the largest fraction of a ship's LDT weight (70–85%), while non-ferrous metals account for another 10%. However, a survey at six recycling yards in China revealed that 99% of materials from dismantled ships can be recovered and, according to material flow analysis, among the materials that can be reprocessed, steel accounts for 74.9% and non-ferrous metals for 7.8% (Steuer et al., 2021).

Plastics are also widely used in marine applications due to their advantageous properties including durability, light weight, corrosion resistance, and cost-effectiveness. Plastics such as HDPE, PVC, PC, PMMA, and glass fiber reinforced polymers are appropriate for boat hulls and decks, windows and weather shields, enclosures, ducts, and containers. In addition, high-performance plastics (e.g., PEEK, PEI, PAI) exhibit excellent strength, stability in high temperatures, and high wear resistance, and are used for bearings, bushings, gears, and engine components (Dimitrellou et al., 2025). The growing interest in replacing metal components with polymer equivalents for weight savings, cost reduction, and environmental benefits makes recycling an urgent necessity, especially for waste from fossil-based plastics, which are difficult to degrade when discarded in the environment.

Addtionally to recycled materials, valuable individual marine items and spare parts are recovered from decommissioned ships, removed, inspected, repaired (if necessary), and reused in ships or similar applications, promoting the CE. Quality tests are carried out to restore original specifications and ensure their reliability, guaranteeing their performance on board. Lower quality items are also sold on the second-hand market and to spare parts suppliers, accompanied by the necessary technical data sheets.

Machinery components usually refer to engine room components such as ferrous and non-ferrous metal pipes, auxiliary engines, shafts, railings, and floor coverings, as well as equipment such as propellers, anchors, chains, mooring winches, and ladders. Typical spare parts include essential components of machinery systems, such as combustion engines, motors, pumps, compressors, boilers, refrigerators, electrical generators, ventilation ductwork, etc.

Electrical and electronic equipment includes various components that can either be recycled for their metal and plastic constituents (e.g., wiring, cables, and circuit boards) or spare parts that can be reused for various purposes (e.g., bridge equipment, communication devices, power distribution panels, lighting equipment, sensors, navigation equipment, generators, batteries, and transformers).

Other items removed from dismantling ships and often sold on the second-hand market include accommodation and office equipment, electrical appliances, kitchen and bathroom fixtures, furnishings and lifesaving equipment (Hiremath et al., 2014). Wooden cladding, wall and ceiling panels – as well as thermal and acoustic insulation of bulkheads, decks, ducts, and pipes – can be recycled or reused, contributing to the CE (Jain et al., 2016).

According to the material flow analysis performed by Steuer et al. (2021), in addition to steel (74.9%) and non-ferrous metals (7.8%), other items that can be reprocessed are ship motors (9.55%), marine equipment (3.15%), furniture (1.55%), wood (0.87%) and wires and cables (0.66%). A slightly different material composition was presented by Jain et al. (2016), based on the lightweight distribution recorded in the stability manual of a 11,000 LTD Handymax bulk carrier. The calculations revealed that ferrous metals acounts for 86%, non-ferrous for 1%, electrical and electronic for 1%, and machinery for 6%. According to the same authors, the developed methodology can provide with a reliable estimation of material streams for an end-of-life ship with an accuracy of 88%.

3.3.2 SWOT Analysis

A strengths, weaknesses, opportunities, and threats (SWOT) analysis was created based on the current situation in an attempt to capture trends, opportunities, potential threats, and constraints in the field (Table 3.1).

Strengths

- **High steel recovery:** Life cycle assessment (LCA) studies show that recycling steel from ships saves a significant amount of primary energy and CO_2 emissions per ton of product compared to primary production, thereby enhancing the economic sustainability of "green steel." This results in competitive prices for recycled steel and a smaller carbon footprint for end users (Rahman et al., 2016).
- **Reuse and remanufacturing of equipment:** The literature emphasizes the importance of dismantling for reuse and remanufacturing, such as with engines, electronics, and piping, which enhances value retention. This reduces capital expenditures (CAPEX) for operators and creates high-value secondary markets (ElMenshawy et al., 2024).

TABLE 3.1
SWOT Analysis

Strengths	Weaknesses
• High steel recovery • Reuse and remanufacturing of equipment • Greater access to green financing • Material traceability	• Compliance costs • Material heterogeneity and hazardous substances • Knowledge and safety gaps • Design for circular disassembly • Unstable demand for second-life parts • Lack of decision support in practice • Reflagging issue
Opportunities	**Threats**
• Blockchain and digital traceability • Collaborations and industrial symbiosis • Green materials markets and certification • Research on decision-making and stochastic models • EU policy and funding • Maritime green transition • Labor health and safety	• The cost advantage of "beaching" method • Volatile scrap prices • Technological risk • Global maritime transportation demand

- **Greater access to green financing:** Adopting circular practices improves transparency and access to green financing, while mitigating the risk of reputational damage from inappropriate practices. CE strategies in shipping are considered a catalyst for differentiation (Okumus et al., 2023).
- **Material traceability:** Mandatory inventory of hazardous materials (IHM) and digital lists of hazardous materials enable the pre-planning of dismantling and safer waste handling, thereby reducing the time taken and the number of accidents (Čulin, 2025).

Weaknesses

- **Compliance costs:** Ship recycling regulations increase investment and operational costs, particularly for smaller dismantlers, thereby creating asymmetrical markets (Alcaide et al. 2017).
- **Material heterogeneity and hazardous substances:** Sealants, paints, PCBs/asbestos, and alloy mixtures make it difficult to separate material streams and increase compliance, environmental, and safety costs (Celik et al., 2024; Mehtaj et al., 2024).

- **Knowledge and safety gaps:** Field research (for instance in Bangladesh) has documented the risks associated with cutting and handling blocks, as well as the inadequate use of personal protective equipment (PPE). These issues undermine productivity and circularity outcomes (Mehtaj et al., 2024).
- **Design for circular disassembly:** The absence of mature, shipbuilding-specific design guidelines results in "afterthought" solutions and missed recovery opportunities. Although recent frameworks emphasize circular design for disassembly, implementation lags behind (Formentini and Ramanujan, 2023; Scipioni et al., 2023).
- **Unstable demand for second-life parts:** Despite the option of remanufacturing, the market often favors new parts due to warranties and standards, which limits cash flow for secondary parts suppliers.
- **Lack of decision support in practice:** Despite the existence of studies on ship-specific recycling plans, few operators make use of real-time digital, multi-criteria optimization tools.
- **Reflagging issue:** Cargo vessels can change their vessel's flag (reflagging) to a flag state that has not entered the Hong Kong Convention, so is alliable to complete the ship recycling process on yards with lower safety and environmental regulations.

Opportunities

- **Blockchain and digital traceability:** Distributed ledger technology (DLT) can verify material and waste routes, certifications, and low-carbon claims, facilitating closed-loop contracts and premium pricing.
- **Collaborations and industrial symbiosis:** Co-productions and contracts between designers, shipyards, scrapyards, and steel mills reduce transaction costs and increase recovery rates (Okumus et al., 2023)
- **Green materials markets and certification:** Certified LCA from scrap enables green material labeling and long-term offtake agreements.
- **Research on decision-making and stochastic models:** Combining LCA with multi-criteria/stochastic methods improves decisions regarding recycling, remanufacturing, and resale.
- **EU policy and funding:** Specific EU funding programs for safe facilities and marine pollution control can support infrastructure upgrades and capacity building.
- **Maritime green transition:** Favorable ship recycling makes it possible to create a new environment-friendly fleet, accelerating the green transition of shipping.

- **Labor health and safety:** Ship recycling in authorized facilities in accordance with HKC 2009 ensures the enhancement of the health and safety of the workers.

Threats

- **The cost advantage of "beaching" method:** The economic advantage of beaching without full safety measures diverts ship traffic away from compliant ship dismantlers, thereby undermining competitive equality.
- **Volatile scrap prices:** Steel demand cycles and macroeconomic fluctuations can reduce recycling yard revenues and encourage the adoption of cheaper practices.
- **Technological risk:** Costs, data governance, and interoperability issues may delay the adoption of blockchain and traceability systems, thereby reducing the expected benefits.
- **Global maritime transportation demand:** Ship recycling must be balanced with the construction of new ships so as not to disrupt the balance of global maritime transport capacity.

The SWOT analysis reveals that the integration of CE principles into ship recycling is influenced by multiple dimensions. Strengths – such as the high value of recovered metals and the existence of regulatory frameworks – suggest that the field already has a solid foundation for further strengthening. At the same time, the prospect of developing new markets for secondary materials is part of a strategy to strengthen European industrial circularity.

However, weaknesses remain decisive. The complexity of international regulatory compliance and the cost of upgrading shipbreaking facilities create significant barriers. Reflagging issues reveal another critical dimension: shipowners who change flags shortly before the end of a ship's life to avoid the strict requirements. This phenomenon undermines the effectiveness of the regulatory framework, creating conditions of unfair competition and limiting the possibilities for the effective implementation of CE principles.

In terms of opportunities, the green transition in the maritime sector offers a powerful means of modernizing shipbuilding and shipbreaking. The shift toward low-emission fuels and decarbonization technologies is linked to the need for sustainable recycling models to ensure the entire life cycle of ships. Furthermore, the emergence of labor health and safety as an opportunity rather than just a risk emphasizes that integrating safety standards can give shipyards a competitive advantage if they invest in modern technology and staff training. Additionally, academia has the opportunity to develop decision-support methodologies that optimize the balance between economic efficiency, environmental compliance, and social responsibility.

Conversely, threats are still being linked to geo-economic and macroeconomic changes. Demand for global maritime transportation creates inherent uncertainty: during periods of high demand, recycling declines as ships remain in operation, while during times of crisis, scrapping increases sharply, often in a way that does not allow for the optimal use of materials. This phenomenon carries the risk of "circularity overload," whereby existing recycling chains are unable to consistently manage large volumes of ships.

3.4 Decision-Making in Ship Recycling

In recent years, ship recycling has captured the interest of many researchers, resulting in the publication of numerous research papers and literature reviews. However, this chapter focuses on decision support methodologies, a research field in which the available literature is very limited. These methodologies are expected to play a contributing role in the optimal adoption of ship recycling in the future.

Despite the abundance of available literature, mainly on ship recycling and thus in the adoption of CE principles, only a limited number of publications incorporate decision support methods in this field of research. The Scopus search engine was selected for the identification of relevant research studies. Using tailored key words, the search has been narrowed to studies including articles or conference papers which apply decision methods in ship recycling. The search yielded an extremely limited number of publications. It is noteworthy that, although no time limit was set on the search period, the identified publications are from 2016 onwards. This obviously explains why there are so few relevant publications (only five research papers), as decision methodologies have been used in other sectors for several years, but their use in ship recycling is very recent and expected to increase in the coming years.

3.4.1 Applied Decision Methods

The most frequently used approach to the modeling sustainable practices in ship recycling is multicriteria analysis. multi-criteria decision-making (MCDM) techniques provide a structure framework for balancing environmental sustainability, economic feasibility and social responsibility. They can assist stakeholders to align with CE principles. During the literature review, the following methodologies were identified as being used.

- **Analytic hierarchy process (AHP)/analytic network process (ANP):** AHP and ANP are widely used methods with applications in various domains. Paired comparisons are made between criteria based on judgments assigned numerical values on a scale from 1–9. ANP is a generalization of AHP, and its basic structures are networks (Strantzali and Aravossis, 2016).

- **The technique for order preference by similarity to ideal solutions (TOPSIS):** The basic concept of this method is that the selected alternative should have the shortest distance from the negative ideal solution in geometrical sense. The ideal alternative has the best level for all criteria, whereas the negative ideal is the one with all the worst criteria values (Strantzali and Aravossis, 2016).
- **Decision-making trial and evaluation laboratory (DEMATEL):** DEMATEL technique uses matrix operations with score and reveal the causal relationships between criteria by categorizing them into cause and effect groups. Causal diagrams are used for visualizing complex causal relationships of criteria and to help to make a deep investigation of the problem (Ozturkoglu et al., 2019).
- **VlseKriterijumska Optimizacija I Kompromisno Resenje (VIKOR):** The method involves ranking alternatives based on their performance across multiple criteria and then identifying the "best compromise" solution that provides the highest overall satisfaction according to a pre-defined set of criteria and weights (Mentes, 2023).
- **Interpretive structural modeling (ISM):** ISM methodology identifies relationships among items. It is based on a structural self-interaction matrix with pairwise comparison of variables. The main limitation of the technique is that a limited number of variables can be considered (Attri et al., 2013).
- **MCDA combined fuzzy methodology:** The classic MCDA methods generally assume that all criteria and their respective weights are expressed in crisp values. Due to the availability and uncertainty of information, sometimes it is difficult to provide exact numerical values for all criteria. In this context, a realistic approach is the use of linguistic variables, applying the combination of MCDA methods and fuzzy set theory (Strantzali and Aravossis 2016).

3.4.2 Decision Criteria

The majority of publications applying decision support methodologies in the field of ship recycling focus on risk assessment decision factors. Ship recycling facilities and processes involve many risks that affect the environment, people and the economy. Worker safety is of the utmost importance, as injuries might be fatal. Furthermore, improper treatment of hazardous materials can cause serious environmental damage. Table 3.2 summarizes the categories of decision factors in the literature (Ozturkoglu et al., 2019; John and Kumar, 2016; Mannan et al., 2022; Mentes, 2023). Four basic categories are

TABLE 3.2

Applied Decision Criteria

Dimension	Factors/Criteria
Environment	Physical and chemical hazards, accidents, green design, the influence of harmful materials
People	Report culture (trust and non-punitive approach, support for staff contributions, psychological safety, rewarding staff for safety-related notifications); awareness (presence and frequency of occupational health and safety trainings, vigilance, and attention; preparedness (resource allocation and redundancy, future-proofing of safety procedures, resilience drills, and feedback); elasticity (flexibility of process/procedure/configuration change capacity, personnel backup working in critical sections, and use of other equipment spares in ship dismantling/recycling operations at the facility), job Stress, job pressure, job burnout, and job autonomy, control, and authority
Economic	Green management, green marketing, green logistics, management commitment, operating and maintenance costs, environmental impact costs
Circular economy	Creating direct and indirect market opportunities, government policies, environmental and maritime regulations, ship recycling yard certification, funding for capacity building, reduction of virgin material liability, decarbonization through ship recycling, infrastructure development, standards and guidelines development

recorded: environment, people, economy and the implementation of the CE are shown separately in Table 3.2 to better reflect the relevant criteria.

The criteria "physical and chemical hazards" and "accidents" reflect the need for integrated risk management in a high-risk environment such as ship dismantling facilities. The inclusion of "green design" indicates a shift toward upstream solutions – i.e., designing ships with the aim of disassembly and safe reuse (design for disassembly). The "influence of harmful materials" factor is linked to the inventory of hazardous materials (IHM) requirements of the Hong Kong Convention and the EU SRR; it is critical for compliance and transparency. Incorporating them into decision-making helps yard managers and regulators prioritize investments in safer dismantling technologies, waste treatment, and infrastructure upgrades.

The "people" dimension reflects the evolution of the approach from "compliance" to "safety culture." "Report culture" and "awareness" refer to modern safety standards whereby non-punitive reporting is key. "Preparedness" and "elasticity" touch on concepts of resilience engineering (e.g., drills, redundancy, process flexibility) that have begun to be applied in high-risk industries. Moreover, psychosocial indicators ("job stress," "burnout," "autonomy," "control") are linked to performance in conditions of high workload and low social protection – they are often underrepresented in technical studies, but their inclusion adds a social dimension to the assessment.

Criteria such as "green management," "green marketing," and "green logistics" show that sustainability is not only a cost but also a strategic advantage (e.g., branding, access to green financing). Operating and maintenance costs and environmental impact costs allow externalities to be incorporated into decisions and LCC (life cycle costing) to be performed. The parallel use of these criteria with MCDM methodologies makes the economic evaluation of investments in technologies and infrastructure more realistic.

"Creating direct and indirect market opportunities" refers to the secondary market for steel and equipment (green steel market). "Government policies" and "environmental and maritime regulations" define the institutional framework (Hong Kong Convention, EU SRR). "Ship recycling yard certification" and "funding for capacity building" indicate the need for institutionalized certification and skills development. "Reduce virgin material liability" and "decarbonization through ship recycling" reflect the contribution of ship recycling to the green transition, while "infrastructure development" and "standards and guidelines development" are linked to the need to systematize practices at the international level.

The selection of criteria such as these creates a multifaceted database that meets the needs of different stakeholders (shipowners, shipbreakers, regulators, and employees) and allows for the creation of decision-support systems that are not one-dimensional but incorporate sustainability and resilience.

3.4.3 Literature Review Discussion

John and Kumar (2016) proposed a decision-making framework for location selection for shipbreaking yards using AHP, considering factors such as cost, legislation, environment, and available labor force. Site selection is a typical multi-criteria problem involving both objective criteria (such as cost and proximity to steel mills) and subjective criteria (such as social impact and political stability). AHP is widely used for such problems because it converts subjective assessments into comparable levels. Transparency in the decision-making process and ease of interpretation of the weights assigned to each criterion are permitted, as is the communication of results to stakeholders (visual pairwise comparisons). However, the AHP method is subject to inconsistencies arising from subjective pairwise comparisons, particularly if the experts involved do not maintain consistent criteria. It also struggles to take uncertainties and stochastic scenarios into account (e.g. changes in scrap prices or regulatory changes). Therefore, AHP is ideal as an initial, transparent site selection tool, facilitating communication between investors and regulatory bodies. However, for strategies involving high uncertainty, sensitivity analysis or stochastic methods are recommended.

Ozturkoglu et al. (2019) proposed a comprehensive preventive risk management framework for the ship recycling industry that covers occupational, environmental, and procedural risks. The framework uses risk prioritization methods and DEMATEL techniques to prioritize areas of intervention.

The field of ship recycling risks is multidimensional, covering safety, pollution, and compliance. The proposed framework enables different types of risk to be combined and preventive measures to be suggested based on priorities and available resources. The advantages of this methodological approach include providing practical guidance for managers, determining which risks to address first and how to allocate resources, and linking risk assessment techniques with management strategy to enhance preventive risk management (PRM). However, some risk metrics can be difficult to measure in organizations with limited monitoring and can be data intensive. Furthermore, implementation in different geographical areas may require contextualization due to the influence of cultural and economic factors on priorities. This framework is particularly well-suited to yards and regulatory authorities looking to transition from a reactive to a proactive approach to risk management.

Mannan et al. (2022) identified critical factors for achieving circularity in ship recycling and proposed a framework based on interpretive structural modeling (ISM) to prioritize and link these factors. The study highlights the importance of government policies, regulatory frameworks, and financing. The use of ISM/MCDM is logical when the goal is to map correlations between multiple, not easily measurable, factors (policies, regulations, financing, expertise). ISM helps to reveal causal hierarchies, leading to comparative weighting and ranking. It enables systematic management of complex, multi-criteria issues and the identification of strategic policy levers. It is ideal for situations whereby there is a wealth of qualitative stakeholder knowledge but limited quantitative data. One disadvantage is that it relies heavily on subjective input from experts, meaning the quality of the results depends on how stakeholders are selected and weighted. It is suitable for the purpose of the article, which is to map factors and propose a strategic framework. However, in order to be transferred into practical yard/owner decisions, it needs to be reinforced with quantitative data and robustness checks.

Mentes (2023) focused on the Aliaga region in Izmir, applying a regional risk analysis that incorporated FMEA, OWGA and FVIKOR techniques to assess the risk profile and resilience of facilities. Field data was collected and advanced MCDM/risk analysis techniques were applied. The combination of FMEA (failure mode effects analysis), OWGA (ordered weighted geometric average), and VIKOR (a non-dominated sorting genetic algorithm) aims to identify potential failure modes in detail and rank them, taking into account asymmetries of importance (OWGA) and a multi-objective approach with compromise (VIKOR). This approach facilitates resilience assessment. It integrates technical details (FMEA) with advanced aggregation techniques (OWGA and Fuzzy VIKOR), which handle different preference and risk profiles. Using real field data increases the reliability of the conclusions. Limitations to the adoption of the methodological framework include high data/expertise requirements, as implementation requires well-documented failure histories and technical knowledge for FMEA scoring. As a complex method, it can make it difficult to communicate the results to non-technical executives. In general, it is highly suitable for large, well-established facilities such as Aliaga, where data and technical expertise are readily available. It is less applicable to smaller or weaker yards without systematic recording.

Celik et al. (2024) conducted a strategic analysis of the Aliaga region, incorporating SWOT, TOWS, and ANP in order to rank strategic interventions and combine internal and external factors, while also allowing for complex interactions between criteria. Integrating SWOT-TOWS with ANP converts qualitative strategic analysis into quantified priorities, enabling strategies that combine internal capabilities with external opportunities and threats to emerge. It successfully converts strategic findings into specific, prioritized actions. The negatives of the method include the complexity of ANP and the need for high-quality pairwise comparisons, which increase the likelihood of errors/subjectivity. There is also the difficulty of translating it into specific technical interventions without additional technical models or costs. It is therefore well-suited to strategic planning at a regional level, such as in Aliaga, where stakeholders need to coordinate policies, funding, and technical interventions.

All these studies recognize the complexity of ship recycling, which is political, technical, economic, and social, and choose MCDM/combined methodologies to address multiple criteria. This approach is appropriate for decisions that are not purely technical, but rather are interdisciplinary in nature. Studies such as those by Mentes (2023) and Ozturkoglu et al. (2019) focus on operational risk and resilience. They use complex, data-rich tools such as FMEA and FVIKOR, which require a technical database. Mannan et al. (2022) and John and Kumar (2016) offer additional strategic and systemic frameworks, such as factor mapping and location selection, and utilize traditional MCDM methods, such as ISM and AHP, which are well-suited to early-stage strategic decisions. Celik et al. (2024) bridge strategic analysis (SWOT/TOWS) with quantification (ANP), ideal for case-based strategic interventions at the regional level.

Shipowners and brokers prefer transparent and fast selection tools, such as AHP and scoring, for local decisions (e.g. reflagging and location). Scrapyard managers and regulatory authorities need technical rigor, such as FMEA and risk frameworks, as well as tools for assessing resilience and compliance (Mentes, 2023; Ozturkoglu et al., 2019). Combining SWOT and ANP is beneficial for policy and strategic planners when it comes to regional planning (Celik et al., 2024).

3.5 Conclusions

Overall, the analysis shows that the transition to a circular model in shipbuilding and ship recycling cannot be viewed in a one-dimensional way. It requires a combination of technological innovations, strict and fair enforcement of regulations, investments in worker health and safety, and the development of academic tools that will act as catalysts for decision-making. The management of these elements will determine whether ship recycling will become a cornerstone of the circular economy (CE) or remain stuck in the contradictions and imbalances of the global shipping system.

Developments in ship recycling are likely to be strongly influenced by international trends in sustainable shipping and the transition to low carbon emissions. The maritime green transition reinforces the need for new ship design models and the use of clean materials from the outset. Thus, the ships of the future will not only be more energy efficient, but also easier to dismantle and recover resources.

Digitalization is expected to play a decisive role, with the implementation of digital "material passports," blockchain, and artificial intelligence for monitoring and tracking materials throughout the life cycle of ships. These technologies can contribute to transparency, reduce the reflagging issue, and strengthen trust among stakeholders.

At the social level, the emphasis on improving the health and safety of workers in shipbreaking yards will be reinforced through stricter regulations and certified training practices. This will not only reduce accidents and occupational diseases, but also increase the attractiveness of facilities that adopt high standards, creating a competitive advantage.

From the perspective of research and academia, there is a strong prospect for the development of multi-criteria decision-making (MCDA) methodologies that combine economic, environmental, and social criteria. Such methodologies can be used by both shipowners and regulatory authorities to optimize recycling strategies.

At the geo-economic level, growing global maritime transportation demand will continue to influence the pace and volume of ship scrapping. However, with the strengthening of international regulations and increasing pressure from financial institutions for green investments, "beaching" practices are expected to gradually give way to more sustainable, technologically advanced solutions.

Overall, the sector is moving toward a hybrid model where the principles of the CE, technological innovation, social responsibility, and global markets will shape the future of ship recycling. Whether this will result in a truly sustainable and resilient practice will depend on the degree of cooperation between all parties involved and their ability to adapt to the challenges ahead.

Furthermore, some common methodological problems and gaps have been identified, as follows.

- **Subjectivity and robustness:** Many MCDM tools rely on expert judgments. These tools require the implementation of sensitivity and uncertainty analyses, as well as cross-validation with real data.

- **Lack of dynamism/stochasticity:** Time-series uncertainty (scrap prices, demand cycles) is rarely incorporated. Tools such as stochastic optimization, real options, or scenario-based MCDA will improve practical value.
- **Data requirements and scaling:** Complex methodologies require high-quality data. Areas with poor data logging, such as some yards in South Asia, are difficult to implement in, as the lack of consistent and reliable datasets prevents proper model calibration, scaling and validation, ultimately reducing the accuracy and applicability of decision-support tools.

Based on the applicability of the examined decision-making approaches, future research could focus on the following.

- **Hybrid frameworks with validation:** A combination of MCDA (AHP/ANP/VIKOR) with stochastic optimization and scenario analysis, which covers both subjective assessments and market uncertainties (e.g., scrap prices and demand).
- **Standardized data templates and digitalization:** In order to reduce subjectivity and enable cross-case validation, we recommend the use of material passports/databases to provide decision tools with quantitative data. This addresses the data requirements issues identified by Ozturkoglu et al. (2019) and Mentes (2023).
- **Layered decision workflows:** The separation could be into three levels, as follows: (a) strategic – SWOT/ANP; (b) local/functional – AHP for location/contract; (c) technical/operational – FMEA + VIKOR for risk mitigation. Each level feeds into the next, providing different types of output/data. This concept is based on the success of combinatorial approaches in the reviewed papers.
- **Validation and policy design:** Policy recommendations should be accompanied by cost/benefit and LCA/LCC assessments, so that regulators can design incentives (e.g., green steel credits) which offset compliance costs. This links the findings of Mannan et al. (2022) and Celik et al. (2024) with applied economic models.

The general consensus is that there is no "one-size-fits-all" approach; the choice of method should depend on the purpose (strategic versus operational), scale (regional versus single yard), and quality of the available data. The most powerful frameworks aim for hybrid designs that emphasize robustness testing, data standardization, and outputs that can be communicated to stakeholders.

References

Alcaide, J.I., Rodríguez-Díaz, E., Piniella, F., 2017. European policies on ship recycling: A stakeholder survey. *Marine Policy*, 81, 262–272

Attri, R., Nikhil Dev, N., Sharma, V., 2013. Interpretive structural modelling (ISM) approach: An overview. *Research Journal of Management Sciences*, 2(2), 3–8. ISSN 2319–1171

Basel Convention 2023 on the Control of Transboundary Movements of Hazardous Wastes and Their Disposal, UN, Revised 2023. chrome-extension://efaidnbmnnnibpcajpcglclefindmkaj/; https://www.basel.int/Portals/4/download.aspx?e=UNEP-CHW-IMPL-CONVTEXT–2023. (Accessed 24 September 2025)

Basel. https://www.basel.int/TheConvention/Overview

BCC Research. https://www.bccresearch.com/market-research/environment/ship-recycling-market.html

BIMCO. https://www.bimco.org/news-insights/trending-topics/ship-recycling/

Celik, M.S., Akyar, D.A., Ceylan, B.O., 2024. Navigating environmental sustainability in ship recycling industry: Integrated SWOT-TOWS-ANP approach. *Ocean Engineering*, 295, 116906

Čulin, J., 2025. Proposal to reclassify SCCPs under international regulations on ship recycling to enhance environmental and human health protection. *Applied Sciences*, 15, 104. https://doi.org/10.3390/app15010104

Decision (EU) 2025/322 amending Implementing Decision (EU) 2016/2323 establishing the European List of ship recycling facilities pursuant to Regulation (EU) No 1257/2013 of the European Parliament and of the Council on ship recycling. https://eur-lex.europa.eu/eli/dec_impl/2025/322/oj/eng (Accessed 24 September 2025)

Dimitrellou, S., Strantzali, E., Iakovidis, I., 2025. A decision-making strategy for selection of FDM-based additively manufactured thermoplastics for industrial applications based on material attributes, *Sustainable Futures*, 9. https://doi.org/10.1016/j.sftr.2025.100640

ElMenshawy, O.M., Ülkü, M.A., Hsuan, J., 2024. Navigating green ship recycling: A systematic review and implications for circularity and sustainable development. *Sustainability*, 16, 7407. https://doi.org/10.3390/su16177407

EMSA, 2008. *Study on the Certification of Ship Recycling Facilities*. chrome-extension://efaidnbmnnnibpcajpcglclefindmkaj/. https://www.american-club.com/files/files/EMSA_Guidance_Inventory_of_Hazardous_Materials_2018.pdf (Accessed 24 September 2025)

EMSA, 2016. *Best Practice Guidance on the Inventory of Hazardous Materials*. https://www.emsa.europa.eu/we-do/sustainability/environment/150-ship-recycling/2874-emsa-s-best-practice-guidance-on-the-inventory-of-hazardous-materials.html (Accessed 24 September 2025)

EMSA, 2017. *Guidance on the Inventory of Hazardous Materials*. https://emsa.europa.eu/about/financial-regulations/items.html?cid=280&id=3003 (Accessed 24 September 2025)

EMSA, 2019. *Guidance Inspections Ship Recycling*. chrome-extension://efaidnbmnnnibpcajpcglclefindmkaj/. https://insb.gr/wp-content/uploads/2024/07/SR_TN_05–2019-EMSA_Guidance_on_Ship_Recycling_PSC_inspections.pdf (Accessed 24 September 2025)

European Commission Waste Shipments. https://environment.ec.europa.eu/topics/waste-and-recycling/waste-shipments_en (Accessed 24 September 2025)

Formentini, G., Ramanujan, D., 2023. Design for circular disassembly: Evaluating the impacts of product end-of-life status on circularity through the parent-action-child model. *Journal of Cleaner Production*, 405, 137009

Geissdoerfer, M., Savaget, P., Bocken, N.M.P., Hultink, E.J., 2017. The circular economy – A new sustainability paradigm? *Journal of Cleaner Production*, 143, 757–768

Hiremath, A.M., Pandey, S.K., Kumar, D., Asolekar, S.R., 2014. Ecological engineering, industrial ecology and eco-industrial networking aspects of ship recycling sector in India. *APCBEE Procedia*, 10, 159–163. https://doi.org/10.1016/j.apcbee.2014.10.035

HKSRC.2/Circ.1, 2024. *Provisional Guidance on the Implementation of Hong Kong and Basel Conventions with Respect to the Transboundary Movement of Ships Intended for Recycling*. https://webaccounts.imo.org/Common/WebLogin.aspx?ReturnUrl=%2f&error_message=login_required (Accessed 24 September 2025)

Hong Kong International Convention for the Safe and Environmentally Sound Recycling of Ships, 2009. https://www.imo.org/en/about/conventions/pages/the-hong-kong-international-convention-for-the-safe-and-environmentally-sound-recycling-of-ships.aspx (Accessed 24 September 2025)

IMO Resolution A.962(23), 2003. *IMO Guidelines on Ship Recycling*. https://webaccounts.imo.org/Common/WebLogin.aspx?ReturnUrl=%2f&error_message=login_required (Accessed 24 September 2025)

IMO Resolution MEPC.178(59), 2009. *Calculation of Recycling Capacity for Meeting the Entry-Into-Force Conditions of the Hong Kong International Convention for the Safe and Environmentally Sound Recycling of Ships*. https://webaccounts.imo.org/Common/WebLogin.aspx?ReturnUrl=%2f&error_message=login_required (Accessed 24 September 2025)

IMO Resolution MEPC.196(62), 2011. *Guidelines for the Development of the Ship Recycling Plan,* adopted by resolution https://webaccounts.imo.org/Common/WebLogin.aspx?ReturnUrl=%2f&error_message=login_required(Accessed 24 September 2025)

IMO Resolution MEPC.197(62), 2011. *Guidelines for the Development of the Inventory of Hazardous Materials*. https://webaccounts.imo.org/Common/WebLogin.aspx?ReturnUrl=%2f&error_message=login_required (Accessed 24 September 2025)

IMO Resolution MEPC.210(63), 2012. *Guidelines for Safe and Environmentally Sound Ship Recycling*, adopted by resolution https://webaccounts.imo.org/Common/WebLogin.aspx?ReturnUrl=%2f&error_message=login_required (Accessed 24 September 2025)

IMO Resolution MEPC.211(63), 2012. *Guidelines for the Authorization of Ship Recycling Facilities*, adopted by resolution https://webaccounts.imo.org/Common/WebLogin.aspx?ReturnUrl=%2f&error_message=login_required (Accessed 24 September 2025)

IMO Resolution MEPC.222(64), 2012. *Guidelines for the Survey and Certification of Ships Under the Hong Kong Convention*, adopted by resolution https://webaccounts.imo.org/Common/WebLogin.aspx?ReturnUrl=%2f&error_message=login_required (Accessed 24 September 2025)

IMO Resolution MEPC.223(64), 2012. *Guidelines for the Inspection of Ships Under the Hong Kong Convention*, adopted by resolution https://webaccounts.imo.org/Common/WebLogin.aspx?ReturnUrl=%2f&error_message=login_required (Accessed 24 September 2025)

IMO Resolution MEPC.379(80), 2023. *Guidelines for the Development of the Inventory of the Hazardous Materials*, adopted by resolution https://webaccounts.imo.org/Common/WebLogin.aspx?ReturnUrl=%2f&error_message=login_required (Accessed 24 September 2025)

IMO Resolution MEPC.405(83), 2025. *Amendments to the 2023 Guidelines for the Development of the Inventory of Hazardous Materials*. https://webaccounts.imo.org/Common/WebLogin.aspx?ReturnUrl=%2f&error_message=login_required (Accessed 24 September 2025)

Jain, K.P., Pruyn, J.F.J., Hopman, J.J., 2013. Critical analysis of the Hong Kong international convention on ship recycling. *Conference Paper October 2013*

Jain, K.P., Pruyn, J.F.J., Hopman, J.J., 2016. Quantitative assessment of material composition of end-of-life ships using onboard documentation, *Resources, Conservation and Recycling*, 107, 1–9. https://doi.org/10.1016/j.resconrec.2015.11.017

John, J., Kumar, S., 2016. A locational decision-making framework for shipbreaking under multiple criteria. *International Journal of Strategic Decision Sciences*, 7(1). https://doi.org/10.4018/IJSDS.2016010104

Lloyd's List, 2025. https://www.lloydslist.com/LL1152167/Ship-recycling-volumes-in-2024-drop-to-lowest-levels-since-2005 (Accessed 24 September 2025)

Mannan, B., Rizvi, Md.J., Dai, Y.M., 2022. Developing a MCDM-based framework for achieving a circular economy through ship recycling. *Proceedings of MARTEC 2022. The International Conference on Marine Technology 21–22 December 2022*, BUET, Dhaka, Bangladesh

Mannan, B., Rizvi, Md.J., Dai, Y.M., 2024. Ship recycling in developing economies of South Asia: Changing liability to a commodity, *Green Technologies and Sustainability*, 2(2). https://doi.org/10.1016/j.grets.2023.100064

Mehtaj, N., Zakaria, NM.G., Awal, Z.I., Dipto, S.S., Hannan, M.A., Dev, A.K., Ali, M.T., 2024. Ship recycling process in Bangladesh and a survey-based risk assessment with mitigation proposal. *Heliyon*, 10(20), e39344. https://doi.org/10.1016/j.heliyon.2024.e39344

Mentes, A, 2023. Risk analysis of on-field and on-board activities and resilience investigation of Izmir Aliaga Ship Recycling Facilities. *Ocean Engineering*, 287, 115891

Milios, L., Beqiri, B., Whalen, K.A., Jelonek, S.H., 2019. Sailing towards a circular economy: Conditions for increased reuse and remanufacturing in the Scandinavian maritime sector. *Journal of Cleaner Production*, 225, 227–235.

Moussa, A.A., Farag, Y.B.A., Gunbeyaz, S.A., Fahim, N.S., Kurt, R.E., 2024. Development and research directions in ship recycling: A systematic literature review with bibliometric analysis. *Marine Pollution Bulletin*, 201, 116247

NGO Shipbreaking Platform. https://shipbreakingplatform.org/spotlight-data-2024/ (Accessed 24 September 2025)

Okumus, D., Gunbeyaz, S.A., Kurt, R.E., Turan, O., 2023. Circular economy approach in the maritime industry: Barriers and the path to sustainability. *Transportation Research Procedia*, 72, 2157–2164

Ozturkoglu, Y., Kazancoglu, Y., Ozkan-Ozen, Y.D., 2019. A sustainable and preventative risk management model for ship recycling industry. *Journal of Cleaner Production*, 238, 117907

Rahman, S.M.M., Handler, R.M., Mayer, A.L., 2016. Life cycle assessment of steel in the ship recycling industry in Bangladesh. *Journal of Cleaner Production*, 135, 963–971

Regulation (EU) 1257/2013 on ship recycling and amending Regulation (EC) No 1013/2006 and Directive 2009/16/EC. https://eur-lex.europa.eu/eli/reg/2013/1257/oj/eng (Accessed 24 September 2025)

Regulation (EU) 2024/1157 on shipments of waste, amending Regulations (EU) No 1257/2013 and (EU) 2020/1056 and repealing Regulation (EC) No 1013/2006. https://eur-lex.europa.eu/eli/reg/2024/1157/oj/eng (Accessed 24 September 2025)

Sarraf, M., Stuer-Lauridsen, F., Dyoulgerov, M., Bloch, R., Wingfield, S., Watkinson, R., 2010. Ship breaking and recycling industry in Bangladesh and Pakistan, Report number: Report No 58275-SAS Affiliation: The World Bank. https://www.researchgate.net/publication/281108572_SHIP_BREAKING_AND_RECYCLING_INDUSTRY_IN_BANGLADESH_AND_PAKISTAN (Accessed 24 September 2025)

Scipioni, S., Dini, G., Niccolini, F., 2023. Exploring circular shipbuilding: A systematic review on circular economy business models and supporting technologies. *Journal of Cleaner Production* 422, 138470

Steuer, B., Staudner, M., Roland, R., 2021. Role and potential of the circular economy in managing end-of-life ships in China. *Resources Conservation and Recycling*, 164. https://doi.org/10.1016/j.resconrec.2020.105039

Strantzali, E., Aravossis, K, 2016. Decision making in renewable energy investments: A review. *Renewable and Sustainable Energy Reviews*, 55, 885–898

Strantzali, E., Livanos, G.A., Aravossis, K., 2023. A comprehensive multicriteria evaluation approach of alterna-2 tive marine fuels. *Energies*, 16, 7498. https://doi.org/10.3390/en16227498

Tola, F., Mosconi, E.M., Marconi, M., Gianvincenzi, M., 2023. Perspectives for the development of a circular economy model to promote ship recycling practices in the European context: A systemic literature review. *Sustainability*, 15, 5919. https://doi.org/10.3390/su15075919

Tsatsaronis, M., 2025. Q2 2025 Report, SHIPPING Market Analysis, National and Kapodistrian University of Athens, 2025. https://container-news.com/q2–2025-shipping-market-analysis/ (Accessed 24 September 2025)

Tukker, A., 2015. Product services for a resource-efficient and circular economy – A review. *Journal of Cleaner Production*, 97, 76–91.

Tuttle, H., 2024. Geopolitical conflict increasing threats to global shipping. *Risk Management*. https://www.rmmagazine.com/articles/article/2024/06/20/geopolitical-conflict-increasing-threats-to-global-shipping (Accessed 24 September 2025)

4

Optimal Network Design of Biowaste Recycling Bins: The Case Study of Cyprus

Pantelis Broukos and Mathaios Panteli

4.1 Introduction

The circular economy (CE) has emerged as a pivotal concept recently in an effort to move towards sustainable use of goods. Based on its fundamental aspects of reducing, reusing, and recycling, the European Commission has established a monitoring framework for CE (Eurostat, 2023), which consists of 11 main indicators structured into five dimensions, as follows.

1. Production and consumption
   a) Material consumption
   b) Green public procurement
   c) Waste generation
2. Waste management
   a) Overall recycling rates
   b) Recycling rates for specific waste streams
3. Secondary raw materials
   a) Contribution of recycled materials to raw-material demand
   b) Trade in recyclable raw materials
4. Competitiveness and innovation
   a) Private investments, jobs, and gross value added
   b) Innovation
5. Global sustainability and resilience
   a) Global sustainability from circular economy
   b) Resilience from circular economy

 DOI: 10.1201/9781003654889-4

The current chapter focuses on the waste management direction and more specifically on the promotion of biowaste recycling. Biowaste refers to *biodegradable waste*, which includes any organic matter in waste that can be broken down into carbon dioxide (CO_2), water, methane, compost, humus, and simple organic molecules by micro-organisms and other living things by composting, aerobic digestion, anaerobic digestion, or similar processes. Biowaste consists mainly of food waste and of organic waste, such as refuse from gardens or parks.

A relatively recent review of papers addressing the recycle bin location/allocation problem in solid waste management can be found in Purkayastha et al. (2015), whose authors present the differences among the studies regarding the adopted model, its objective function, its assumptions, and the used data. The review depicts the variety of the studies, with the model ranging from heuristic methods to exact ones, with or without the use of geographic information system (GIS) data, minimization of transportation costs, maximization of serviced population, etc.

An interesting study is Anghinolfi et al. (2013), who propose a dynamic optimization model for solid waste recycling. The study focuses on modelling the daily quantity of waste in each bin, whose location is already predefined, while determining the daily quantity of material collected in the area and the collecting vehicles routes. The developed decision model is integrated in a GIS-based decision support system (DSS) and applied in the Cogoleto municipality in Italy.

In 2015, Boskovic and Jovicic (2015) introduced a heuristic approach to the optimal relocation of waste bins. Various parameters were considered, such as street directions, spatial location of objects and number of inhabitants, existing location of waste bins, and radius of their coverage. The methodology was implemented in a district in the central area of the city of Kragujevac in Serbia and resulted in reducing the number of installed bins, while offering the same quality of service. This led to a reduction of waste collection cost, and to economic and environmental benefits.

On the contrary, instead of applying a heuristic, mixed integer linear programming (MILP) models are introduced in Bennekrouf et al. (2020) for the solution of the bin location problem. The models take into account maximal coverage distance of each type of bin, which is addressed through GIS data. The models are applied in the city of Boudjlida in Tlemcen in Algeria, with promising results offering a useful decision tool to the stakeholders.

Another very interesting study is Yousefloo et al. (2023), whereby a holistic network approach is undertaken such that a multi-objective scenario-based robust stochastic optimization model for designing a sustainable municipal solid waste (MSW) management network under uncertainty is proposed. The problem is addressed as a six-echelon network (waste generation points, waste hybrid separation centers, transfer stations, replacement truck warehouses, treatment centers, and customers) with various flows from waste generation to recycled energy and materials. The authors emphasize fuel consumption, which affects transportation costs and CO_2 emissions, and apply their model on a real case study of Qazvin province in Iran.

Similarly, in Boresta et al. (2024), a MILP is proposed to solve the strategic planning of treatment and disposal facilities of municipal waste. The model considers vehicle allocation to shifts and driver requirements, as well as vehicle maintenance issues, while evaluating location options for depots, waste treatment facilities, and transshipment points. It is applied to the metropolitan city of Rome in Italy and results in a significant reduction in distance traveled and decreased transit time.

Focusing on the recycling of biowaste, Santos and Mendes (2024) examine rural municipalities with a strong forestry component and addresses vehicle route optimization for the selective collection and transportation of biowaste. While not developing a mathematical model, the authors use the network analyst extension of the ArcGIS software package for travel time route optimization, optimal location selection, and definition of service areas, applying their methodology on the Oleiros-Amieira region in Portugal.

Unlike the existing literature, this chapter focuses on the strategic network design of biowaste bins in a country-sized area. More specifically, a novel mathematical model has been developed that optimizes the placement of the bins, as a facility location problem (FLP). The proposed MILP includes constraints capacity and budget constraints and maximizes the kilograms of CO_2 equivalent that are saved based on the recycled biowaste quantity. Additionally, a methodology is suggested for the geographical estimation of the volumes of biowaste generation, which is inserted to the formulation as input data. Finally, a sensitivity analysis of the optimal solution based on different values of the available budget is conducted. The main contribution of the chapter is that it offers a realistic and visualized tool that can help stakeholders make optimal decisions as to where to place biowaste bins and how many of them to place. The proposed model is generic and can be easily applied on the network design of other infrastructure, as well.

The remainder of the chapter is organized as follows. In Section 4.2, the proposed generic mathematical model is introduced, together with the preprocessing procedure for its application on Cyprus. In Section 4.3, the case study to which the methodology is applied, together with the computational results, are presented. Finally, in Section 4.4, the chapter is concluded and suggestions are made for future research.

4.2 Methods and Analysis

4.2.1 Mathematical Formulation

As previously mentioned, the strategic design of biowaste bins network is formulated as an FLP. The notation used in the mathematical model is presented in Table 4.1.

TABLE 4.1

Nomenclature

Sets and Indices	
I	Set of areas under study, $I = \{1,2,...,n\}$
K	Set of different types of bins to be established (for example, bins with different capacity)
J	Set of major districts to which the areas under study belong, $j = \{1,2,...,m\}$
i	Index for areas under study, $i \in I$
k	Index for the type of bin to be established, $k \in \mathcal{K}$
j	Index for the major districts, $j \in J$
Parameters	
$\overline{PPLT_i}$	Population in area i
$\overline{DSTRCT_{ij}}$	Parameter that takes the value of 1 if the area i belongs to the major district j and 0 otherwise
$\overline{D_i}$	Biowaste generation of area i (kg)
$\overline{A_i}$	Area of demand area i (km^2)
$\overline{P_k}$	Capacity of bin of type k (lt)
$\overline{F_{ik}}$	Already existing bins of type k in demand area i
$\overline{CV_k}$	Variable cost of installing a bin of type k (€)
$\overline{CF_k}$	Fixed cost of installing at least one bin of type k in an area per square kilometer (€/km^2)
$\overline{BDG}$	Total available budget (€)
$\overline{SC}$	Safety coefficient, so that the system does not reach maximum capacity (for example, 0.7)
$\overline{E}$	CO_2 equivalent that is saved per biowaste recycled (kg of CO_2/kg of biowaste recycled)
$\overline{BigM}$	A very large number
Decision Variables	
x_{ik}	General integer variable that is equal to the number of bins of type k that are installed at area i
y_i	Continuous variable that is equal to the biowaste that is eventually recycled at area i
z_{ik}	Binary variable that is equal to 1 if at least one bin of type k is installed at area i, and 0 otherwise

Thus, the FLP is formulated as follows.

$$\text{Maximize} \sum_{i \in I} \bar{E} \times y_i \tag{4.1}$$

Subject to the following constraints:

$$\sum_{k \in K} \sum_{i \in I} \overline{CF_k} \times \bar{A}_t \times z_{ik} + \overline{CV_k} \times x_{ik} \leq \overline{BDG} \tag{4.2}$$

$$y_i \leq \overline{SC} \times \sum_{k \in K} \overline{P_k} \times (x_{ik} + \overline{F_{ik}}), \forall i \in I \tag{4.3}$$

$$y_i \leq \bar{D}_{t,} \forall i \in I \tag{4.4}$$

$$z_{ik} \leq x_{ik} \leq \overline{BigM} \times z_{ik} \forall i \in I, k \in K \tag{4.5}$$

$$x_{ik} \in \mathbb{Z}^{+} \forall i \in I, k \in K \tag{4.6}$$

$$y_i \in \mathbb{R}^{+} \forall i \in I \tag{4.7}$$

$$z_{ik} \in \{0,1\} \forall i \in I, k \in K \tag{4.8}$$

The objective function (4.1) expresses the maximization of the total CO_2 equivalent that is saved, after the establishment of the biowaste recycling bins network. The constraint (4.2) guarantees that the total (fixed and variable) cost of establishing the network will not exceed the total available budget. The constraints (4.3) express that the biowaste quantity that is eventually recycled at area i cannot be more than the total capacity of the bins installed at that area, taking into consideration the safety coefficient. The constraints (4.4) assure that the biowaste quantity that is eventually recycled at area i cannot be more than the actual biowaste quantity generated at that area. The constraints (4.5) connect the decision variables x_{ik} and z_{ik}. Constraints (4.6)–(4.8) limit the values of the decision variables.

4.2.2 Preprocessing Procedure

As can be noticed, the previously described mathematical model could be easily translated in a generic form. This characteristic offers to the model the ability to be easily adaptable to solve any FLP. In the herein application, the biowaste bin location problem is addressed, and the correspondence of the parameters between the two problems, the generic and the specific one, is depicted in Table 4.2.

The geographical area where the model is to be applied can be discretized into smaller areas; for each one, the number of every type of

TABLE 4.2

Correspondence of Parameters of the Generic Facility Location Model with the Ones in the Biowaste Bin Location Model

A/A	Biowaste Bin Location Model	Generic Facility Location Model
1	Biowaste Bins	Facilities
2	Capacity of the biowaste bin, in liters	Capacity
3	Types of biowaste bins with different capacity	Types of facilities
4	Biowaste generation in area i, in kilograms	Demand level of demand area i

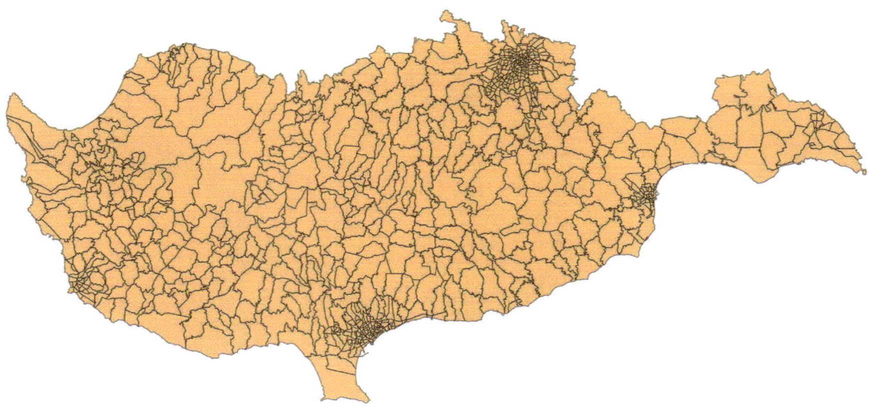

FIGURE 4.1

Postal domains in Cyprus depicted in QGIS software.

biowaste bins to be installed will be derived from the solution. Here, the area under study is Cyprus and for the spatial direction of the problem, the actual map of the whole Cyprus is considered. To this end, real GIS data are used and Cyprus is divided using the open-source software QGIS into areas based on the postal domains (National Open Data Portal, 2024), as shown in Figure 4.1. Each of these 855 postal domains, apart from its post code, includes a geographical code as well. Thus, every postal domain belongs to a major district of Cyprus, as shown in Table 4.3.

Finally, for each of these domains, the actual population is used based on the population census of 2021 (Cyprus Statistical Service, 2024). Moreover, the area of every postal domain is calculated in km^2 and its population density can be derived, as shown in Figure 4.2.

TABLE 4.3

Correspondence Between the Geographical and Post Codes of a Postal Domain and the District to Which It Belongs

A/A	District	Geographical Code	Post Code
1	Lefkosia	1XXX	1XXX, 2XXX
2	Ammochostos	3XXX	5XXX
3	Larnaka	4XXX	6XXX, 7XXX
4	Lemesos	5XXX	3XXX, 4XXX
5	Pafos	6XXX	8XXX

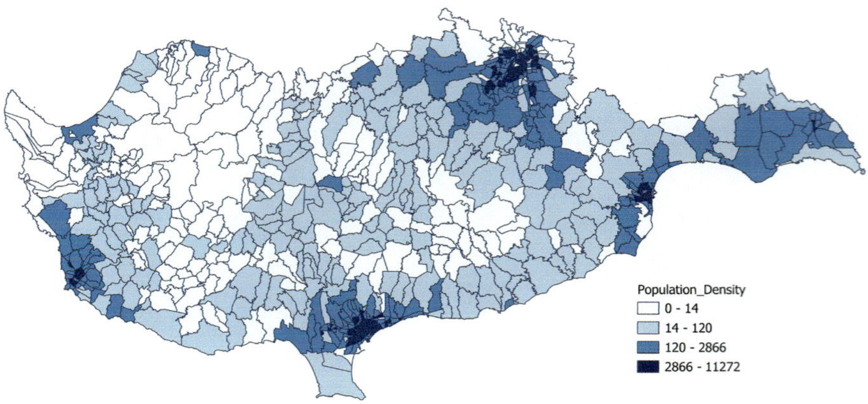

FIGURE 4.2
Population density of the postal domains in Cyprus. The colors of the heat map are explained at the bottom right of the figure (inhabitants/km^2).

4.3 Results

4.3.1 Case Study

For the estimation of biowaste generation in every domain, the amount of food waste generated per year divided by the average population of the country is considered. Food waste is measured as fresh mass all along the food value chain, including production sites, processing and manufacture, retail and distribution, restaurants and food services, and households. Based on the Eurostat data (Eurostat, 2023), the annual food waste generation in Cyprus is estimated at 397 kg per capita.

The previously described model exports the total capacity of biowaste bins to be installed in order to recycle the generated biowaste. For this reason, it is assumed that the bins will be emptied once every week, and therefore, their capacity should be enough to hold the biowaste generated weekly, i.e., 397 × 7/365 = 7.61 kg/capita. Thus, for every domain the generated biowaste is computed as $\overline{D_i} = 7.61 \times \overline{POP_i}$, since its population is already known.

Regarding the parameter $\overline{E}$ of the model, the European Commission's Joint Research Centre, in the report "End-of-Waste Criteria for Biodegradable Waste Subjected to Biological Treatment (Compost & Digestate)" (European Commission's Joint Research Centre, 2014), provides an assessment which indicates that anaerobic digestion of biowaste, followed by biogas utilization, can save around 0.5–1.0 kg of CO_2 equivalent per kilogram of biowaste when considering both methane capture and replacement of fossil fuels. For this reason, in order to make a moderate estimate, the parameter $\overline{E} = 0.5\, kg\ of\ CO_2\ equivalent\ /\ kg\ of\ biowaste\ recycled$.

Moreover, in the current study, two types of bins are considered as shown in Figure 4.3. Also, their characteristics are depicted in Table 4.4.

Additionally, the previously described safety coefficient was chosen to be $\overline{SC} = 0.7$. This means that if at a region the generated biowaste is 70 liters, then a bin capacity of 100 liters should be installed in order to recycle it. This so called "safety coefficient" is inserted so that the system does not reach maximum capacity. In other words, if a 70-liter bin was installed to collect 70 liters of biowaste, then the respective bin might be always full and may some days overflow.

4.3.2 Results: Sensitivity Analysis

The previously described model was coded in Python. The model was solved using Google's open-source optimization software OR-TOOLS. More specifically, the solver SCIP is used, which is the default OR-TOOLS mixed integer programming solver. The instances were solved on a PC with Windows 10 (64-bit), Intel Core i5: 4200M CPU @ 2.50GHz and 8Gb RAM.

As it was already mentioned, different values were chosen for the available budget to invest in order to conduct sensitivity analysis (€1 million, €2 million, €3 million, €4 million, €5 million, €6 million, €7 million). For each instance, the optimal number of biowaste bins in every postal domain in Cyprus is calculated.

Specifically, Figure 4.4 shows the number of biowaste bins with capacity of 240 liters to be installed in every postal domain in Cyprus for a budget of €3 million. Similarly, the optimal solution for a budget of €5 million is shown in Figure 4.5. As it can be easily noticed by comparing Figure 4.4 with Figure 4.5, if more money is invested on creating the network, then more areas would be covered and more biowaste bins would be installed.

If this expected result is extended, then one could say that the bins to be installed cannot continue to grow in number if more and more money is invested. There must be a "turning point" at which the number of installed

FIGURE 4.3
Types of bins with different capacity: (a) 240 liters, (b) 110 liters.

TABLE 4.4

Characteristics of the Biowaste Bins

Type *k*	Bin Capacity	Bin Variable Cost	Bin Fixed Cost
1	240 liters	€100	€1,000
2	1,100 liters	€500	€5,000

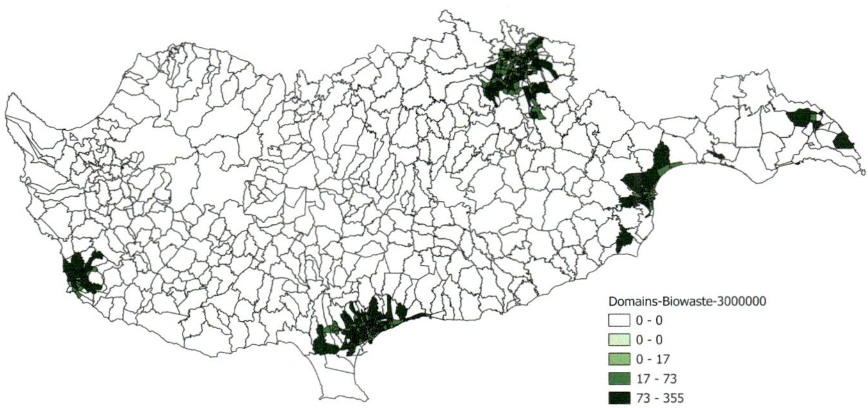

FIGURE 4.4
Number of biowaste bins with capacity 240 liters to be installed in every postal domain in QGIS software for a budget of €3 million. The colors of the heat map are explained at the bottom right of the figure.

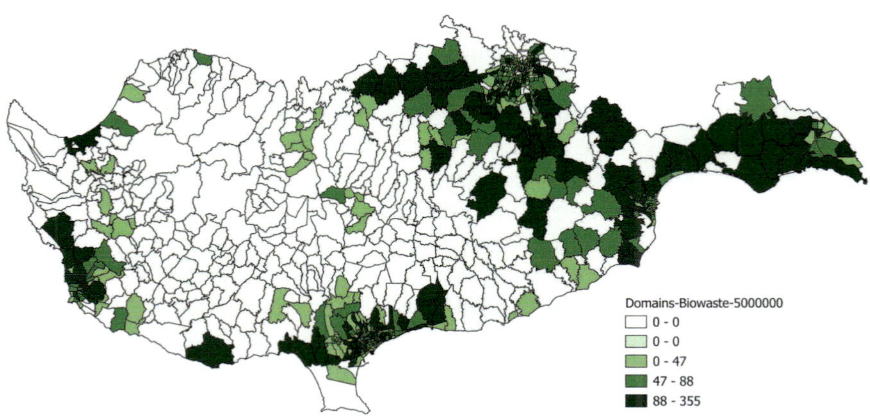

FIGURE 4.5
Number of biowaste bins with capacity 240 liters to be installed in every postal domain in QGIS software for a budget of €5 million. The colors of the heat map are explained at the bottom right of the figure.

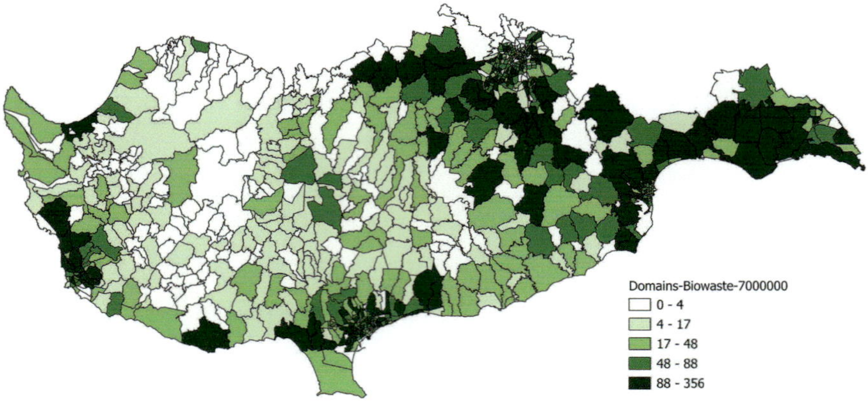

FIGURE 4.6
Number of biowaste bins with capacity 240 liters to be installed in every postal domain in QGIS software for a budget of €5 million. The colors of the heat map are explained at the bottom right of the figure.

bins can hold all the generated biowaste quantity. Having a budget that exceeds this "turning point" would alter neither the quantity of biowaste recycled nor the CO_2 equivalent of emissions saved. In the current study, this "turning point" is €7 million, as will be later shown.

Figure 4.6 depicts the number of biowaste bins with capacity 240 liters to be installed in every postal domain in Cyprus for a budget of €7 million. By comparing Figure 4.6 with Figure 4.4 and Figure 4.5, it is easily noticed that with a total available budget of €7 million, biowaste bins would be installed in even more areas.

At this point, it has to be mentioned that the numbers of the total available budget are strongly dependent on the rest of the parameters of the problem, such as the cost and the capacity of every type of bin. This means that the sensitivity analysis can be even more extended to incorporate combinations of different values for different parameters of the problem. However, this extension is out of the scope of this study.

The optimal number of biowaste bins in every district of Cyprus, if the higher level of the postal domains is examined, is depicted in Figure 4.7 for different values of available budget. It has to noted, that in the optimal solutions for every value of available budget, only bins of 240 liters capacity (Figure 4.3a) and no bins of 1,100 liters capacity (Figure 4.3b) are chosen, which is mainly because the former offers better cost per capacity rate (Table 4.4).

Furthermore, moving on to the higher level from the districts to the whole country and taking into consideration that for every value of the available

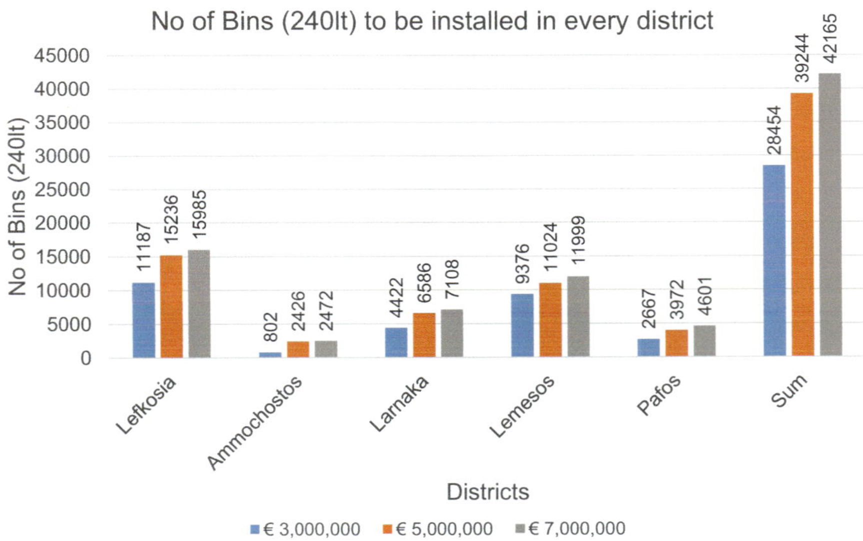

FIGURE 4.7
Number of biowaste bins (240 liters) to be installed in every district of Cyprus for different values of available budget.

budget, a different number of biowaste bins are installed, this leads to different total capacity, and therefore, a different quantity of biowaste that can be recycled. However, the weekly biowaste generated quantity is assumed to be stable and deterministic. This means that there is certainly a value of the available budget for which the biowaste recycled is equal to the biowaste generated, i.e., all generated biowaste is recycled. As mentioned earlier, increasing the budget more than this value has no impact on the total recycled quantity and the total savings of CO_2 emissions. In order to detect the size of this budget, a close look at the comparison of the generated and the recycled quantity for every instance is taken.

Thus, for all instances, the weekly generated biowaste is the same and equal to 7,021,049.4 kg, based on its calculation presented in Section 3.1. According to the model's results, when € million is invested, then 1,650,094.9 kg of biowaste are recycled, i.e., 23.5% of the total generated quantity. Similarly, when €3 million is invested, 4,778,766.4 kgs of biowaste are recycled, i.e., 68.1% of the total generated quantity.

Following this reasoning, Figure 4.8 shows the quantities of biowaste that will be recycled in every district and totally in the whole country for the different values of available budget for which the sensitivity analysis was conducted. As it can be easily noticed, investing €7 million leads to an optimal network of biowaste bins that can collect the total generated quantity. This

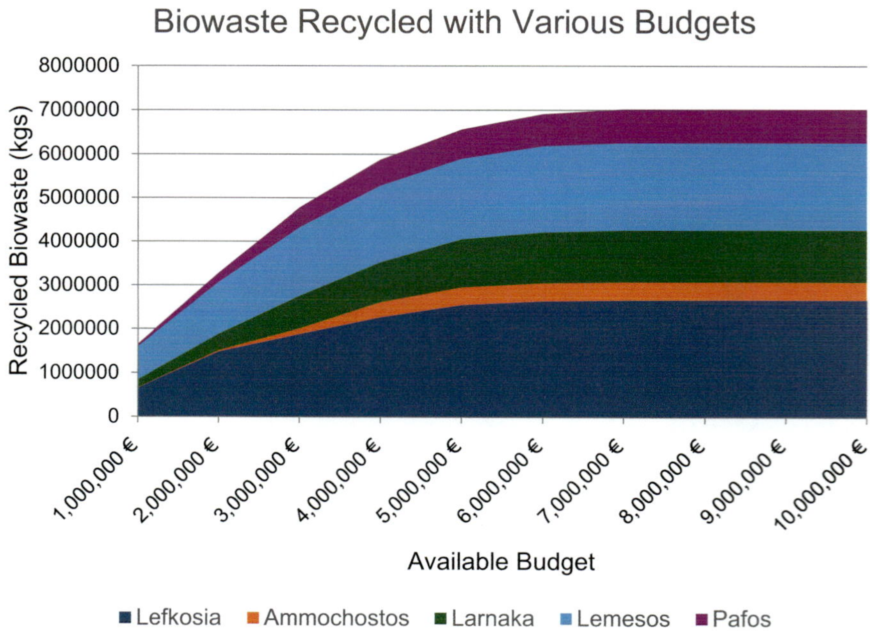

FIGURE 4.8
Biowaste recycled in each district in Cyprus for different values of available budget for the bins installation.

means that investing more than €7 million will not lead to recycling more biowaste and will increase the network's cost without any benefit. Finally, another useful insight derived from Figure 4.8 and the sensitivity analysis is that the stakeholders can gradually increase the budget to invest in order to achieve different levels of recycling. For example, if they do not wish to invest €7 million from the beginning, then they can start by investing €1 million, for example. Then, after several months or years, they can invest another €1 million to expand the network of biowaste bins and collect more biowaste. This can gradually continue until €7 million is invested and all generated biowaste is recycled.

Finally, another important result derived from the optimal solution of the proposed mathematical model is the calculation of the CO_2 equivalent that would be saved if the respective biowaste quantity were recycled. Based on the formula presented in Section 3.1, the CO_2 equivalent is linearly proportional to the recycled biowaste quantity. Thus, Table 4.5 summarizes the results of the sensitivity analysis including the kilograms of CO_2 equivalent that will be saved after the installation of the biowaste recycling network.

TABLE 4.5

Summary of Results, Including CO_2 Saved

A/A	Available Budget	Biowaste Generated (kg)	Biowaste Recycled (kg)	Biowaste Recycled Percentage (%)	CO_2 Saved (kg)
1	€1 million	70,21049.4	1,650,094.9	23.5%	825,047.5
2	€2 million	70,21049.4	3,276,977.0	46.7%	1,638,488.5
3	€3 million	70,21049.4	4,778,766.4	68.1%	2,389,383.2
4	€4 million	70,21049.4	5,882,285.3	83.8%	2,941,142.6
5	€5 million	70,21049.4	6,571,028.1	93.6%	3,285,514.1
6	€6 million	70,21049.4	6,913,825.7	98.5%	3,456,912.9
7	€ million	70,21049.4	7,020,088.6	100.0%	3,510,044.3
8	€8 million	70,21049.4	7,020,976.3	100.0%	3,510,488.2
9	€9 million	70,21049.4	7,021,049.4	100.0%	3,510,524.7
10	€10 million	70,21049.4	7,021,049.4	100.0%	3,510,524.7

4.4 Discussion: Conclusion

The optimal network design of biowaste bins is addressed in this chapter. To this end, a generic and adaptable mathematical model is proposed and an easily applicable geographical discretization is suggested. The formulation is applied on Cyprus, for which geographic information systems (GIS) data are gathered, processed, and inserted to the model. The results of the model are also depicted in GIS format for better visualization. The integration of spatial, economic, and environmental data ensures that the proposed solutions are both practical and sustainable. A sensitivity analysis is conducted and the overall developed methodology is a powerful tool for decision-makers and stakeholders because it provides a rigorous, data-driven approach that can help guide investment and policy decisions. Moreover, it offers flexibility, allowing stakeholders to make informed decisions based on available resources and future projections.

Furthermore, future research should focus on a sophisticated procedure for more realistic and accurate estimation of the generated biowaste. Regarding this estimation, a sensitivity analysis can be conducted in order to gain optimal solutions with different generated biowaste quantities. Finally, an extension of the current work could be the study and modeling of the next steps of the circular economy of recycled biowaste, which can be energy production, fertilizer creation, and so on.

4.4.1 Acknowledgments

This research is funded by the European Union's Horizon 2020 program, which is being held at KOIOS Research and Innovation Center of Excellence,

Department of Electrical and Computer Engineering, University of Cyprus, and with a partnership with the United Nations Educational, Scientific and Cultural Organization (UNESCO) in the context of the corresponding author's secondment period as Chair on Green Innovation and Circular Economy, School of Mechanical Engineering, National Technical University of Athens, Greece.

References

Anghinolfi, D., M. Paolucci, M. Robba and A. C. Taramasso, "A dynamic optimization model for solid waste recycling," *Waste Management,* vol. 33, pp. 287–296, 2013.

Bennekrouf, M., W. Aggoune-Mtalaa, K. Benladghem and H. O. Cherif, "A strategic approach for the optimal location of recycling bins in the city of Boudjlida in Algeria," in *2020 IEEE 13th International Colloquium of Logistics and Supply Chain Management (LOGISTIQUA),* Morocco, 2020.

Boresta, M., A. L. Croella, C. Gentile, L. Palagi, D. M. Pinto, G. Stecca and P. Ventura, "Optimal network design for municipal waste management: Application to the metropolitan city of rome," *Logistics,* vol. 8, no. 79, 2024.

Boskovic, G. and N. Jovicic, "Fast methodology to design the optimal collection point locations and number of waste bins: A case study," *Waste Management & Research,* vol. 33, no. 12, pp. 1094–1102, 2015.

Cyprus Statistical Service, "Population – Predefined Tables," 2024. [Online]. Available: https://www.cystat.gov.cy/en/KeyFiguresList?s=46.

European Commission's Joint Research Centre, *End-of-Waste Criteria for Biodegradable Waste Subjected to Biological Treatment (Compost & Digestate): Technical Proposals,* European Commission, Seville, 2014.

Eurostat, "Monitoring framework," 2023. [Online]. Available: https://ec.europa.eu/eurostat/web/circular-economy/monitoring-framework.

National Open Data Portal, "Postal Domain Boundaries," 2024. [Online]. Available: https://data.gov.cy/el/dataset/oria-tahydromikon-tomeon-ktimatikos-horometrikos-hartis.

Purkayastha, D., M. Majumder and S. Chakrabarti, "Collection and recycle bin location-allocation problem in solid waste management: A review," *Pollution,* vol. 1, no. 2, pp. 175–191, 2015.

Santos, B. and C. Mendes, "GIS-based approach for optimizing biowaste collection services in rural small sized municipalities," *Waste Management Bulletin,* vol. 2, no. 1, pp. 9–20, 2024.

Yousefloo, A., R. Babazadeh, M. Mohammadi, A. Pirayesh and A. Dolgui, "Design of a robust waste recycling network integrating social and environmental pillars of sustainability," *Computers & Industrial Engineering,* vol. 176, p. 108970, 2023.

5

EV Infrastructure Optimization to Support EU Climate Objectives

Pantelis Broukos and Mathaios Panteli

5.1 Introduction

Nowadays, the transportation sector is responsible for a substantial contribution to greenhouse gas emissions (GHGs). In an effort to reduce this contribution, electric vehicles (EVs) are promoted, while investments are made for greener electrical energy production mainly through the integration of renewable energy sources (RES).

An interesting study on the optimal allocation of charging stations is Baouche et al. (2014), whose authors estimate the energy needs by exploiting origin/destination (OD) information. The problem is modeled by a modified fixed charge location model, which is applied on a dense urban network in the Lyon metropolitan area of France.

On the contrary, in Mohsenzadeh et al. (2015), a genetic algorithm is used to solve the problem of optimal siting and sizing of EV public charging stations. The authors approach the problem from the scope of power system reliability and city traffic layout and attempt to minimize both the energy not supplied (ENS) and the charging station's related costs. Their methodology is applied on a simulated radial distribution network.

Another important study is (Lee and Madanat, 2017) introduces a bi-criteria model minimizing both GHGs and users' costs. The authors consider several constraints, such as limited available budget, station accessibility and peak-hour energy demands. The problem is modeled as a convex program and it is implemented on an arbitrarily generated urban network. The conducted sensitivity analysis shows that the higher the available budget, the bigger the GHG emissions reduction.

Moreover, in Faridimehr et al. (2019), a two-stage stochastic programming model is proposed in order to determine the optimal network of charging stations for a community. The authors introduce a sophisticated approach which considers several parameters, such as uncertainties in the arrival time

DOI: 10.1201/9781003654889-5

and parking duration of vehicles, the state of charge of arriving vehicles' batteries, drivers' walking ranges and charging preferences, demand during weekdays and weekends, and the community's rate of EV adoption. They implement the sample average approximation method and propose an effective heuristic so as to solve large-scale problems and produce near-optimal solutions. An extension this work is Fazeli et al. (2021), whose authors apply the L-shaped decomposition method instead of the previously developed heuristic, while both studies are applied on the midtown area of Detroit, Michigan, United States.

Furthermore, in order to identify the optimal location of charging stations for EVs in a road transport network, Fredriksson et al. (2019) present an iterative approximation technique to solve the so-called route node coverage (RNC) problem. This technique aims to identify the most probable routes in a transportation network and the minimum number of charging stations and their locations, so that any EV in the transportation network is able to reach a charging station before fully depleting its energy source. The methodology is applied on the well-known Sioux Falls network (South Dakota, USA) and the southern Sweden road network, offering useful insights.

Another interesting work is Liu et al. (2023), who address the charging station location problem (CSLP) and reformulate it as a reinforcement learning problem. The authors introduce a novel model which incorporates an attention layer into the proximal policy optimization (PPO) algorithm, enhancing the algorithm's capacity to identify and understand the intricate interdependencies between different nodes in the network. Their approach considers the balance in charging station coverage across the network and the total time spent during the charging process and is applied in real-world road networks in Stanford and Culver City in California (United States), Queenstown (Singapore), Cambridge (UK) and Rouen (France).

Reinforcement learning is also applied in Heo and Chang (2024), which focuses on the optimal locations of EV fast charging stations (EVFCS). More specifically, the authors introduce a robust decision-making model for optimal EVFCS placement planning integrated with solar power supply, utilizing an advantage actor-to-critic (A2C) deep reinforcement learning (DRL) approach. The methodology handles large-scale problems with 1,000 charging stations and is applied on the complex urban environment of Chicago (United States).

Similar to the current work, Wang et al. (2024) studies the optimal location of electric chargers in a statewide framework. Instead of focusing on a specific urban area, the authors utilize a real-time transportation and charging station dataset in Texas (United States) in order to identify optimal charger placement strategies to minimize travel time by examining the congestion and charging time trade-offs.

The current chapter focuses on the strategic network design of public electric chargers in the area under study. More specifically, a novel mathematical model has been developed that optimizes the placement of the chargers in countrywide framework, as a facility location problem (FLP). The proposed mixed-integer linear problem (MILP) includes power and budget constraints and maximizes the kilograms of carbon dioxide (CO_2) equivalent that are saved through the installation of chargers, which promote the use of EVs. Finally, a sensitivity analysis of the optimal solution based on different values of the available budget is conducted. The main contribution of the chapter is that it offers a realistic and visualized tool that can help stakeholders make optimal decisions as to where to place electric chargers and how many of them to place. The proposed model is generic and can be easily applied on the network design of other infrastructure, as well.

The remainder of the chapter is organized as follows. In Section 5.2, the proposed generic mathematical model is introduced, together with the preprocessing procedure for its application on Cyprus. In Section 5.3, the case study at which the methodology is applied, together with the computational results, are presented. Finally, in Section 5.4, the chapter is concluded and suggestions are made for future research.

5.2 Methods and Analysis

5.2.1 Mathematical Formulation

As previously mentioned, the strategic design of an electric chargers network is formulated as an FLP. The notation used in the mathematical model is presented in Table 5.1.

Thus, the FLP is formulated as follows. **Maximize** $\sum_{i \in I} \overline{E} \times y_i$ (5.1)

This equation is subject to the following constraints.

$$\sum_{k \in K} \sum_{i \in I} \overline{CF_k} \times z_{ik} + \overline{CV_k} \times x_{ik} \leq \overline{BDG} \quad 5.2)$$

$$y_i \leq \overline{SC} \times \sum_{k \in K} \overline{P_k} \times \left(x_{ik} + \overline{F_{ik}} \right), \forall i \in I \quad (5.3)$$

$$y_i \leq \overline{D_i}, \forall i \in I \quad (5.4)$$

$$z_{ik} \leq x_{ik} \leq \overline{BigM} \times z_{ik}, \forall i \in I, k \in K \quad (5.5)$$

$$x_{ik} \in \mathbb{Z}^{+} \forall i \in I, k \in K \quad (5.6)$$

$$y_i \in \mathbb{R}^{+} \forall i \in I \quad (5.7)$$

$$z_{ik} \in \{0,1\} \forall i \in I, k \in K \quad (5.8)$$

TABLE 5.1

Nomenclature

Sets and Indices	
I	Set of demand areas, $I=\{1,2,\ldots,n\}$.
K	Set of different types of chargers to be established (for example, chargers with different kilowatt ratings)
J	Set of major districts to which demand areas belong, $j=\{1,2,\ldots,m\}$
i	Index for demand area, $i \in I$
k	Index for a type of charger to be established, $k \in K$
j	Index for a major district, $j \in J$
Parameters	
$\overline{PPLT_i}$	Population in demand area i
$\overline{DSTRCT_{ij}}$	Parameter that takes the value of 1 if the demand area i belongs to the major district j, and 0 otherwise
$\overline{D_i}$	Power demand level of demand area i
$\overline{P_k}$	Power capacity of charger of type k
$\overline{F_{ik}}$	Already existing chargers of type k in demand area i
$\overline{CV_k}$	Variable cost of establishing a charger of type k (€)
$\overline{CF_k}$	Fixed cost of establishing at least one charger of type k in a demand area (€)
$\overline{BDG}$	Total available budget (€)
$\overline{SC}$	Safety coefficient, so that the system does not reach maximum capacity (for example, 0.7)
$\overline{E}$	CO_2 equivalent that is saved per power demand level (kg of CO_2 per kW of power met)
$\overline{BigM}$	A very large number
Decision Variables	
x_{ik}	General integer variable that is equal to the number of chargers of type k that are established at area i
y_i	Continuous variable that is equal to the power demand that is eventually met at area i
z_{ik}	Binary variable that is equal to 1 if at least one charger of type k is established at area i, and 0 otherwise

The objective function (5.1) expresses the maximization of the total CO_2 equivalent that is saved, after the establishment of the public electric chargers network. The constraint (5.2) guarantees that the total (fixed and variable) cost of establishing the network will not exceed the total available budget.

TABLE 5.2

Correspondence of Parameters of the Generic Facility Location Model with the Ones in Electric Chargers Location Model

A/A	Generic Facility Location Model	Electric Chargers Location Model
1	Facilities	Electric chargers
2	Capacity	Power of the electric charger, in kilowatts
3	Types of facilities	Types of electric chargers with different power
4	Demand level of demand area i	Power needed to charge EVs in area i, in kilowatts

TABLE 5.3

Correspondence Between the Geographic and Post Codes of a Postal Domain and the District to Which It Belongs

A/A	District	Geographic Code	Post Code
1	Lefkosia	1XXX	1XXX, 2XXX
2	Ammochostos	3XXX	5XXX
3	Larnaka	4XXX	6XXX, 7XXX
4	Lemesos	5XXX	3XXX, 4XXX
5	Pafos	6XXX	8XXX

The constraints (5.3) express that the power demand that is eventually met at area i cannot be more than the total power capacity of the chargers established at that area, taking into consideration the safety coefficient. The constraints (5.4) assure that the power demand that is eventually met at area i cannot be more than the actual power demand of that area. The constraints (5.5) connect the decision variables x_{ik} and z_{ik}. Constraints (5.6)–(5.8) limit the values of the decision variables.

5.2.2 Preprocessing Procedure

As it can be noticed, the previously described mathematical model could be easily translated in a generic form. This characteristic offers to the model the ability to be easily adaptable to solve any FLP. In the herein application the electric chargers location problem is addressed, and in Table 5.2, the correspondence of the parameters between the two problems – the generic one and the specific one – is depicted.

The geographic area at which the model is to be applied can be discretized into smaller areas; for each one, the number of every type of electric chargers to be installed will be derived from the solution. Here, the area under study is Cyprus, and for the spatial direction of the problem, the actual map of the whole of Cyprus is considered. To this end, real geographical information system (GIS) data are used and Cyprus is divided in demand areas based on the postal domains (National Open Data Portal, 2024). Also, each of these 855 postal domains, apart from the post code, includes a geographic code, as well. Thus, every postal domain belongs to a major district of Cyprus, as shown in **Error! Reference source not found.** For each of these sectors, the actual population is used based on the population census of 2021 (Cyprus Statistical Service, 2024).

5.3 Results

5.3.1 Case Study

For the application of the problem on Cyprus, five types of public electric chargers, whose characteristics are depicted in Table 5.4, were considered.

The new network to be installed is designed to be able to support 10,000 new EVs in Cyprus. In order to examine how many public electric chargers should be installed in every postal domain, the number of new EVs in every postal domain should be first calculated. To this end, the 10,000 new EVs should first be distributed in the five districts of Cyprus. In order to estimate this, the percentage distribution of the 1,313 newly registered EVs in 2023 is used (Republic of Cyprus, 2024), as shown in the Table 5.5.

However, the goal is to distribute the EVs on all postal domains. For this reason, for every district, the population distribution among the postal domains of the district is used as a weight factor in order to distribute the estimated EVs of the district (Table 5.5, last column) among its domains. In order to illustrate this assumed distribution of the 10,000 new EVs among the postal domains of Cyprus, Figure 5.1 depicts the considered EV density, which is derived by dividing the estimated number of EVs in every postal domain by its area in square kilometers.

Moreover, this estimated number of EVs should be interpreted in terms of the energy needed in every postal domain. To achieve this, the following

TABLE 5.4

Characteristics of the Public Electric Chargers Considered

Type k	Charger Capacity (kW)	Charger Variable Cost	Charger Fixed Cost
1	22	€2,000	€5,000
2	50	€25,000	€30,000
3	100	€35,000	€80,000
4	150	€70,000	€110,000
5	320	€150,000	€200,000

computation is made. The average capacity of an electric battery in an EV is around 40kWh, but can get some up to 100kWh (E.ON Energy Solutions Limited, 2024). So by assuming that every vehicle has an average capacity of around 70kWh and it is needed to charge in one hour, then a charger of 70 kW is needed for every vehicle. However, if we assume that the EVs of a sector are charged in public chargers once a week, then for every vehicle, 10 kW of installed electric chargers are needed.

As regards the amount of the emissions of CO_2 which are saved per kilowatt (kW) of installed capacity for EV chargers, this depends on several factors, including the energy mix of the electricity used, the efficiency of the EV, and the CO_2 emissions of the internal combustion engine (ICE) vehicles being replaced. However, a general figure can be estimated based on average values.

TABLE 5.5

Distribution of EVs Among the Five Districts of Cyprus

A/A	District	EVs Registered in 2023 (Republic of Cyprus, 2024)	Percentage of EVs Registered in 2023 (Republic of Cyprus, 2024)	Estimated New EVs
1	Lefkosia	491	37.4%	3740
2	Ammochostos	113	8.6%	861
3	Larnaka	144	11.0%	1097
4	Lemesos	453	34.5%	3450
5	Pafos	112	8.5%	853
SUM		**1313**	**100.0%**	**10,000**

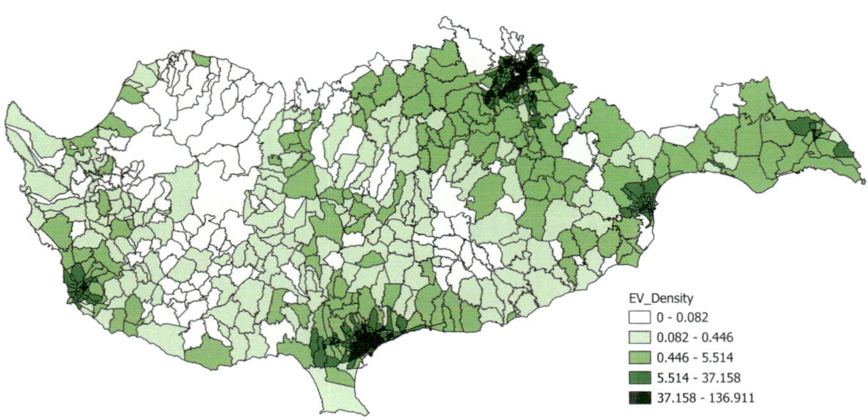

FIGURE 5.1
Considered electric vehicle density among the postal domains in Cyprus (electric vehicles/km²).

TABLE 5.6

Calculation of Carbon Intensity of the Electricity Produced in Cyprus

Type of Energy Source	Share (International Energy Agency, 2022)	Carbon Intensity ($g\ CO_2$ / k Wh) (Ajanovic and Haas, 2019)	Share × Carbon Intensity ($g\ CO_2$ / k Wh)
Coal	1.4%	1,000	14
Oil	85.1%	820	697.82
Wind, solar, etc.	6.6%	30	1.98
Biofuels and waste	6.9%	40	2.76
Average	**100.0%**	**Sum**	**716.56**

According to Department for Energy Security and Net Zero (2024), a medium passenger ICE Vehicle running on petrol emits approximately 177 g of CO_2 per kilometer.

EV emissions vary significantly depending on local electricity generation mix and vehicle usage patterns. According to International Energy Agency (2022), the mix of the energy supply in Cyprus is coal (1.4%); oil (85.1%); wind, solar, etc. (6.6%); and biofuels and waste (6.9%). These numbers show that the energy supply in Cyprus primarily relies on fossil fuels for electricity generation, particularly oil, with a growing but still relatively small share of renewables (solar and wind).

In order to determine the carbon emissions, each type of electricity generation is converted in a certain amount of CO_2 per kWh. To achieve the amount of CO_2, the carbon intensity of electricity generation was determined for each type of fuel that is used to produce electricity, through the work of Ajanovic and Haas (2019). Table 5.6 explains the calculation made to compute the overall carbon intensity of the electricity produced in Cyprus. Thus, the estimated CO_2 emissions per kWh of electricity generated in Cyprus are 716.56 g CO_2 per kWh.

As mentioned earlier, ICE vehicles typically emit about 177 g CO_2 per kilometer. With the Cyprus energy mix emitting about 716.56 g CO_2 per kWh, and assuming EVs consume about 15–20 kWh per 100 km (average 17.5 kWh per 100 km), the emissions would be:

$$EVCO_2\ emissions\ per\ km = \frac{716.56g\,CO_2\ /\ kWh \times 0.175\,kWh\ /\ km}{1km} \approx 125gCO_2\ /\ km \tag{5.9}$$

Therefore, the savings of CO_2 emissions per km = $CO_2$from ICE – CO_2 from EV. Using these numbers, the savings would be 177 g CO_2/km – 125 g CO_2/km = 52 g CO_2/km

Moreover, it is assumed that a 1 kW charger operates for two hours a day, delivering 730 kWh annually. If the EV travels 5 km per kWh (typical for efficient EVs), this corresponds to 3,650 km of travel annually, so the total CO_2 savings per kW of installed charger capacity would be $3{,}650\ \text{km} \times 52\ \text{g}\ CO_2/\text{km} = 189{,}800\ \text{g}\ CO_2 = 189.8\ \text{kg}\ CO_2$ saved per kW (of installed chargers) per year.

5.3.2 Results

The previously described model was coded in Python. The model was solved using Google's open-source optimization software OR-TOOLS. More specifically, the solver SCIP is used, which is the default OR-TOOLS mixed integer programming solver. The instances were solved on a PC with Windows 10 (64-bit), Intel Core i5: 4200M CPU @ 2.50GHz and 8Gb RAM.

As already mentioned, different values were chosen for the available budget in order to conduct sensitivity analysis (€5 million, €7.5 million, €10 million, €12.5 million, €15 million, €17.5 million, €20 million, €25 million). For each instance, the optimal number of public electric chargers in every postal domain in Cyprus is calculated. For example, for a budget of €5 million, Figure 5.2 shows the number of chargers with power 22kW to be installed in every postal domain in Cyprus. Another example is the optimal solution for a budget of €12.5 million, as shown in Figure 5.3. As can be easily noticed by comparing Figure 5.2 with Figure 5.3, the more funding which is invested on creating the network, the more areas are covered and the more chargers are installed.

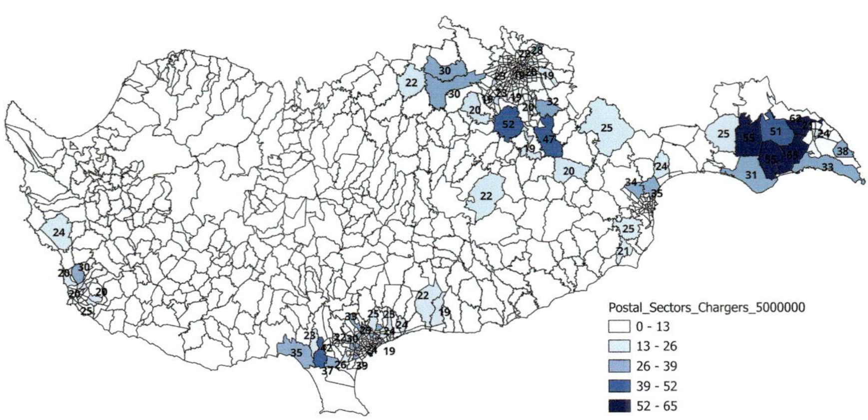

FIGURE 5.2
Number of public electric chargers with power 22kW to be installed in every postal domain in QGIS software for a budget of €5 million. The colors of the heat map are explained at the bottom right of the figure.

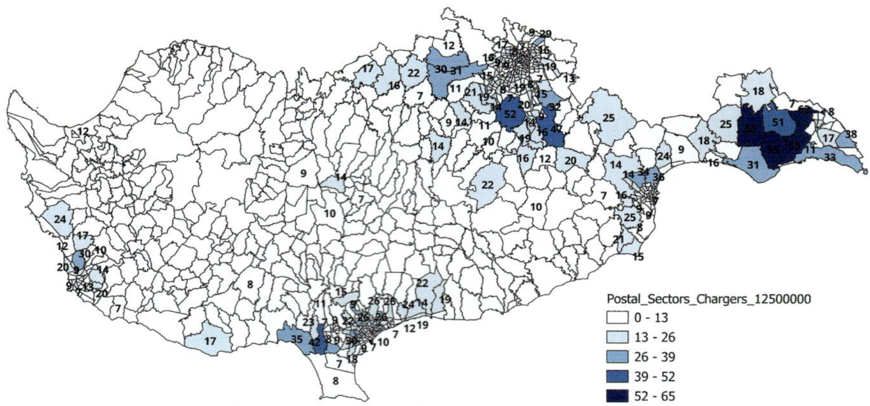

FIGURE 5.3
Number of public electric chargers with power 22kW to be installed in every postal domain in QGIS software for a budget of €12.5 million. The colors of the heat map are explained at the bottom right of the figure.

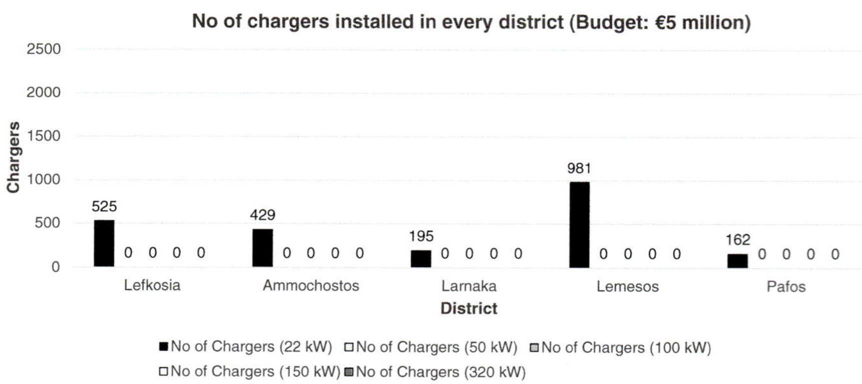

FIGURE 5.4
Number of electric chargers to be installed in every district of Cyprus with an available budget of €5 million.

Moreover, if the higher level of the postal domains is examined, then the optimal number of electric chargers in every district of Cyprus is depicted at Figure 5.4 and Figure 5.5 for a budget of €5 million and 12,5 million, respectively.

Furthermore, moving on to the higher level from the districts to the whole country – and taking into consideration that for every value of the available budget, different numbers of public electric chargers are installed – leads to different total power capacity. However, the daily power needed is assumed to be stable and deterministic. This means that there is certainly a value of the available budget for which the total installed power is equal to the power needed. Increasing the budget more than this value has no impact on the

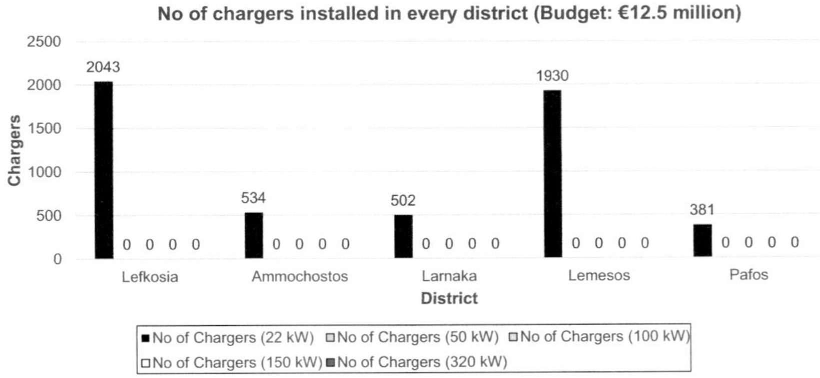

FIGURE 5.5
Number of electric chargers to be installed in every district of Cyprus with an available budget of €12.5 million.

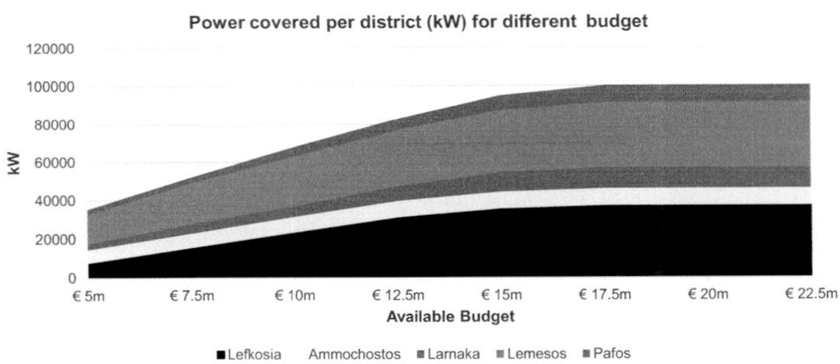

FIGURE 5.6
Power covered in each district in Cyprus for different values of available budget for the installation of chargers.

total power that is delivered and the total savings of CO_2 emissions. In order to detect the size of this budget, a close look at the comparison of the needed and the installed power for every instance is taken.

Thus, for all instances, the daily needed power for the 10,000 new EVs is the same and equal to 100,000 kW, based on its calculation presented in Section 3.1. For example, according to the model's results, when €5 million is invested, 35,290.8 kW of power is delivered to the EVs, i.e., 35.3% of the total generated quantity. Similarly, when €12.5 million is invested, 82,802.9 kW of power is delivered to the EVs, i.e., 82.8% of the total power needed.

Following this reasoning, Figure 5.6 shows the amount of needed energy that will be finally covered, in every district and totally in the whole country, for the different values of available budget, for which the sensitivity analysis was conducted. As it can be easily noticed, investing €17.5 million leads to an optimal network of electric chargers that can cover the total needed power. This means that investing more than €17.5 million will not lead to covering more power and will increase the network's cost without any benefit. Finally, another useful insight derived from Figure 5.6 and the sensitivity analysis is that stakeholders can gradually increase the budget to invest in order to achieve different levels of coverage. For example, they can start by investing €5 million, for example. Then, after several months/years, they can invest another €2.5 million to expand the network of chargers and cover more EVs. This can gradually continue until €17.5 million is invested and all needed power is covered.

5.4 Discussion – Conclusion and Future Work

The optimal network design of public electric vehicle (EV) chargers is addressed in this chapter. To this end, a generic and adaptable mathematical model is proposed and an easily applicable geographical discretization is suggested. The formulation is applied on Cyprus, for which geographic information systems (GIS) data are gathered, processed, and inserted into the model. The results of the model are also depicted in GIS format for better visualization. The integration of spatial, economic, and environmental data ensures that the proposed solutions are both practical and sustainable. A sensitivity analysis is conducted, and the overall developed methodology is a powerful tool for decision-makers and stakeholders by providing a rigorous, data-driven approach that can help guide investment and policy decisions. Moreover, it offers flexibility, allowing stakeholders to make informed decisions based on available resources and future projections.

Furthermore, future research should focus on a sophisticated procedure for more realistic and accurate estimation of the power needed to support the integration of the EVs. Regarding this estimation, a sensitivity analysis can be conducted in order to gain optimal solutions with different needs in power. Finally, an extension of the current work could be the geographical incorporation of the existing chargers into the case study of Cyprus in order to detect how they affect the optimal solution.

5.4.1 Acknowledgments

This research is funded by the European Union's Horizon 2020 program, which is being held at KOIOS Research and Innovation Center of Excellence, Department of Electrical and Computer Engineering, University of Cyprus, and with a partnership (in the context of the corresponding author's secondment period) with the United Nations Educational, Scientific and Cultural Organization (UNESCO) Chair on Green Innovation and Circular Economy, School of Mechanical Engineering, National Technical University of Athens, Greece.

References

Ajanovic, A. and R. Haas, "On the environmental benignity of electric vehicles," *Journal of Sustainable Development of Energy, Water and Environment Systems,* vol. 7, no. 3, pp. 416–431, 2019.

Baouche, F., R. Billot, R. Trigui and N.-E. El Faouzi, "Efficient allocation of electric vehicles charging stations: optimization model and application to a dense urban network," *IEEE Intelligent Transportation Systems Magazine,* vol. 6, no. 3, pp. 33–43, 2014.

Cyprus Statistical Service, "Population – predefined tables," 2024. [Online]. Available: https://www.cystat.gov.cy/en/KeyFiguresList?s=46.

Department for Energy Security and Net Zero, "Greenhouse gas reporting: conversion factors," Department for Energy Security and Net Zero, London, UK, 2024.

E.ON Energy Solutions Limited, 2024. [Online]. Available: https://www.eonnext.com/electric-vehicles/costs-and-benefits-of-electric-cars/battery-capacity-and-lifespan#:~:text=What%20is%20the%20average%20battery,before%20you%20need%20to%20recharge.

European Council, "Fit for 55 package," 2023. [Online]. Available: https://www.consilium.europa.eu/en/policies/fit-for-55/.

Faridimehr, S., S. Venkatachalam and R. B. Chinnam, "A stochastic programming approach for electric vehicle charging network design," *IEEE Transactions on Intelligent Transportation Systems,* vol. 20, no. 5, pp. 1870–1882, 2019.

Fazeli, S. S., S. Venkatachalam, R. B. Chinnam and A. Murat, "Two-stage stochastic choice modeling approach for electric vehicle charging station network design in urban communities," *IEEE Transactions on Intelligent Transportation Systems,* vol. 22, no. 5, pp. 3038–3053, 2021.

Fredriksson, H., M. Dahl and J. Holmgren, "Optimal placement of charging stations for electric vehicles in large-scale transportation networks," *Procedia Computer Science,* vol. 160, pp. 77–84, 2019.

Heo, J. and S. Chang, "Optimal planning for electric vehicle fast charging stations placements in a city scale using an advantage actor-critic deep reinforcement learning and geospatial analysis," *Sustainable Cities and Society,* vol. 113, p. 105567, 2024.

International Energy Agency, "Energy system of cyprus," 2022. [Online]. Available: https://www.iea.org/countries/cyprus.

Lee, J. and S. Madanat, "Optimal design of electric vehicle public charging system in an urban network for greenhouse gas emission and cost minimization," *Transportation Research Part C: Emerging Technologies,* vol. 85, pp. 494–508, 2017.

Liu, J., J. Sun and X. Qi, "Optimal placement of charging stations in road networks: a reinforcement learning approach with attention mechanism," *Applied Sciences,* vol. 13, no. 14, p. 8473, 2023.

Mohsenzadeh, A., C. Pang, S. Pazouki and M. Haghifam, "Optimal siting and sizing of electric vehicle public charging stations considering smart distribution network reliability," in *2015 North American Power Symposium (NAPS),* Charlotte, NC, USA, 2015.

National Open Data Portal, "Postal domain boundaries," 2024. [Online]. Available: https://data.gov.cy/el/dataset/oria-tahydromikon-tomeon-ktimatikos-horometrikos-hartis.

Republic of Cyprus, "Registration of motor vehicles, January–December 2023," Statistical Service of Cyprus, 2024.

Wang, J., H. D. Kaushik and J. Zhang, "Optimal planning of electric vehicle charging stations: integrating public charging networks and transportation congestion," in *IEEE Kansas Power and Energy Conference (KPEC),* Manhattan, KS, 2024.

6

Hybrid Renewable Energy System Analysis for Addressing Water and Electricity Needs on the Island of Mykonos

Athanasios-Foivos Papathanasiou, Ioannis Platanitis, and Evangelos Baltas

6.1 Introduction

Water is a vital resource, influenced by both natural processes and human activities, such as urbanization, agriculture, and industry (Wulandari et al. 2024). Climate change further affects its availability by reducing soil absorption, increasing pollution, and depleting groundwater reserves (Fulazzaky et al. 2023). Population growth intensifies water demand across sectors, placing additional pressure on resources and aquifers (Papathanasiou et al. 2025a). Effective water management, through hydraulic works and conservation technologies, is essential to address such challenges (Bertsiou and Baltas 2022).

Renewable energy sources (RES) play a crucial role in sustainable water management, particularly for energy-intensive processes like desalination (Lagaros et al. 2024). On Greek islands, water scarcity is often aggravated by tourism development and groundwater over-extraction (Alexakis 2022). The use of solar and wind energy to power desalination plants offers a sustainable solution to water shortages, while also contributing to the reduction of greenhouse gas emissions (Papathanasiou and Baltas 2025).

Integrated water and energy management, supported by RES, is vital for adapting to climate change (Skroufouta et al. 2024). Strategic investments and targeted policies can transform Greek islands into models of green development, ensuring the sustainability of water and energy resources, strengthening resilience to environmental challenges, and improving the quality of life (Kalogiannidis et al. 2023).

Overall, the interdependence between water and energy highlights the importance of integrated approaches for sustainable island development. Energy is required for water treatment, pumping, and desalination, while

DOI: 10.1201/9781003654889-6

water is essential for energy production, storage, and cooling processes. The mutual relationship, commonly described as the water–energy nexus, becomes even more critical in island regions where natural resources are limited and climate change increases seasonal pressures. The design of systems that simultaneously address both sectors, islands can reduce reliance on fossil fuels, enhance resilience to extreme events, and ensure the long-term sustainability of essential services for local communities and tourism-driven economies (Papathanasiou et al. 2025b).

Relevant studies have examined various hybrid renewable energy system (HRES) configurations. For instance, Qudah et al. (2025) evaluated a hybrid solar–wind system with water-tank storage for reverse osmosis desalination, identifying reduced water costs down to 0.95 $/m^3$ and 5–12 tons annual carbon dioxide (CO_2) savings, highlighting improved techno-economic performance. Tsiaras et al. (2025) evaluated off-grid electricity pathways for Nisyros Island using wind turbines and photovoltaic panels, identifying three wind turbines as the lowest-cost way to meet demand, highlighting wind-leaning hybrids as economically viable routes to electricity autonomy. Papathanasiou et al. (2024) analyzed an HRES on Skyros Island, comparing pumped-storage hydropower, hydrogen storage, and a combined approach, finding that all scenarios fully cover water needs, while electricity coverage reaches 88% with pumped storage, 83% with hydrogen, and 95% with the combined option. Mendez-Morales et al. (2025) proposed a modular floating energy island for Crete that integrates offshore wind turbines and solar panels, identify an eastern offshore site using resource and environmental criteria, and outline storage and grid-connection needs, pointing at a practical path to energy independence and surplus export. Kiehbadroudinezhad et al. (2023) demonstrated that a wind turbine plus battery system powering seawater reverse osmosis has the lowest environmental impact, about 0.143 kg of carbon dioxide per m^3 and roughly 90% lower damages than diesel desalination. Babatunde et al. (2022) evaluate hybrid energy options for a Nigerian university, identifying a solar–diesel–battery system as the most cost-effective, with 139 $/MWh, 71% renewable contribution, and about 108 tons CO_2 annual savings compared to diesel generation.

Despite the growing literature on HRESs, research focusing on islands with intense seasonal tourism, such as Mykonos, remains limited. The strong fluctuations in population and resource demand create unique challenges that differ from other islands with more stable consumption profiles. Furthermore, most of the existing studies emphasize either energy autonomy or water supply security, without fully integrating both dimensions under varying priority scenarios. The present research work addresses this gap by analyzing an HRES for Mykonos, combining wind energy, desalination, and pumped-hydro storage. The objective is to evaluate alternative management priorities, energy first and water first, to identify the most effective pathway to simultaneously secure water and electricity needs, while enhancing resilience and sustainability on a highly touristic Aegean island.

The novelty of the present work lies in the comprehensive integration of water and energy management priorities within a single hybrid renewable energy framework. Unlike previous studies that predominantly focus on either electricity autonomy or water supply security, the present analysis explicitly tests two distinct operational scenarios in order to capture the trade-offs among energy coverage, water availability, and storage performance. In addition, Mykonos represents a unique case study due to its exceptionally high seasonal variability, driven by tourism, which imposes extreme fluctuations on both electricity and water demand. The combination of desalination powered by renewable energy, large-scale pumped-hydro storage, and scenario-based management strategies provides new insights into how HRESs can be tailored to islands with unstable consumption patterns. This approach not only advances the methodological framework for HRESs but also offers practical guidance for policymakers and engineers seeking resilient solutions for islands facing similar pressures.

6.2 Study Area: Methodology

Mykonos is an island in the Aegean Sea, located in Greece, and belongs to the Cyclades island group (Figure 6.1). The capital of the island is called Chora, situated on its western side. According to the Hellenic Statistical Authority Census of 2021, Mykonos had a population of 10,704 inhabitants (ELSTAT 2021). The island covers an area of 86 km^2 and has a coastline length of 89 km. The port of Mykonos lies approximately 117 km from the port of Piraeus, the main port of Greece. Nearby islands are Tinos (9 km), Syros (29 km), Paros (28 km), and Naxos (26 km).

Mykonos, as a study area of the HRES, is of particular interest since it shows strong seasonal variations in the number of visitors. With a permanent population of 10,704 inhabitants, the island hosts more than two million tourists over a period of about six months (excluding cruise ship passengers, who add around 800,000 additional "visitors)." This means that the island's demand for water and electricity shows significant and sudden changes during the year.

The main source of water supply in Mykonos is groundwater extracted from wells and submarine springs. However, due to the decline in rainfall and the increase in consumption, the island's groundwater resources are not sufficient to meet demand. This has led to seawater intrusion in several wells, resulting in the degradation of groundwater quality.

According to the Municipal Water Supply and Sewerage Company of Mykonos, the maximum daily water consumption reaches 14,000 m^3. With distribution losses estimated at 20–30%, this corresponds to 2,800–4,200 m^3 of water lost per day during the summer months (DEYAM 2025). The total annual water demand on Mykonos is estimated at 4,218,250 m^3.

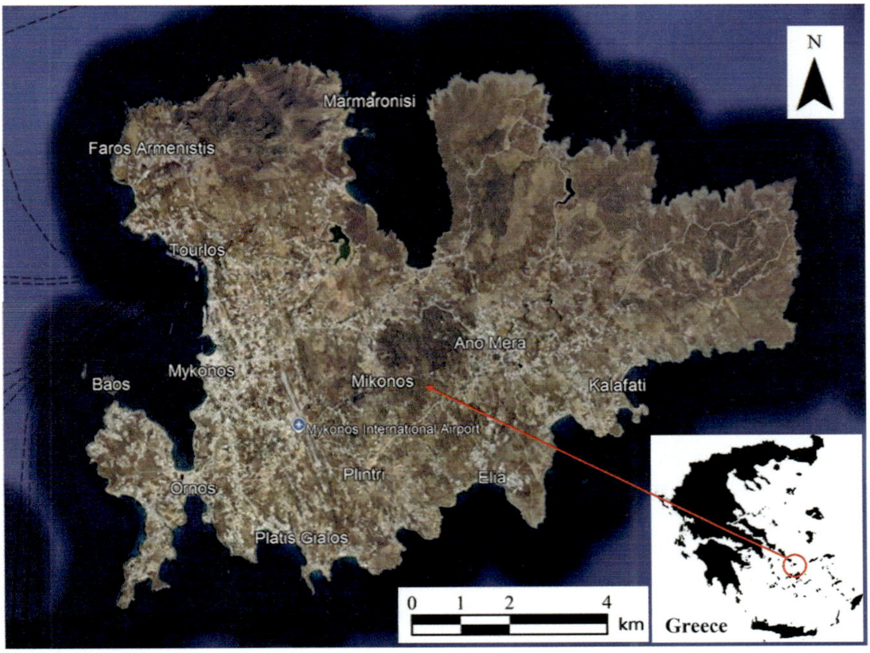

FIGURE 6.1
Map of Mykonos and the island's relative position in Greece.

Source: Image created by the authors based on Google

The energy needs of Mykonos are mainly covered through the Interconnected System of the Cyclades, which links the island to the mainland grid of Greece and secures electricity transfer during peak demand, especially in summer. Tourism development has significantly increased consumption from hotels, restaurants, and other facilities, making reliable energy supply essential for sustaining the island's economy and image (Economou 2010). In addition, Mykonos has a 55 MW oil-fired power plant that operates as backup during high demand or system failures. However, reliance on imported fuel is costly and environmentally burdensome, requiring around 30 shipments per year. Overall, annual electricity demand reaches 264,062 MWh, underlining the importance of energy autonomy and efficient resource management.

In the examined HRES, wind turbines generate electricity, desalination units produce water to meet the island's demand and pumped-hydro storage is used to store excess energy.

The HRES in Mykonos consists of 22 wind turbines, each with a capacity of 2.3 MW (total 50.6 MW). The selection of 22 turbines is based on an energy output corresponding to 1.5 times of the total electricity demand of the island. The calculation of the wind energy produced by the wind turbines was based on wind speed data and the manufacturer's power curve. The wind speeds recorded at the meteorological station are adjusted to match the height at which the wind turbines operate. Specifically, considering the altitude of the meteorological station (13 m) and the hub height of the wind turbines (138 m), the wind data time series is converted accordingly.

To meet water needs, the system includes four desalination units with a total capacity of 1,400 m^3/hour, with an estimated energy consumption of 3.5 kWh/m^3 and a total power demand of 4.5 MW. In addition, two pumps with a combined power of 60 kW are included to transfer the desalinated water.

For the pumped-hydro storage system, four pumps with a total capacity of 2.2 MW are used to transfer seawater into two reservoirs with a combined storage volume of 3,900,000 m^3, located 40 m above sea level. In addition, the system includes hydropower stations with discharge capacities of 20 m^3/s and 40 m^3/s, situated downstream of the Ano Mera and Marathio dams, respectively.

The calculation of daily water demand for Mykonos is based on monthly population data and the assumption that each person consumes an average of 250 liters of water per day. Electricity demand is estimated using monthly population data and an average per capita consumption of 2.5–3.5 kWh/day, accounting for daily activities and the higher use of appliances during the summer tourist season.

Two scenarios were developed for the study of the HRES. In the first scenario, priority is given to meeting the island's electricity demand. Surplus energy is then stored through pumped-hydro storage and finally used for desalination. In the second scenario, priority is given to water needs (desalination), followed by energy storage, and electricity demand is covered last, if surplus energy remains.

6.3 Results

The coverage rates and quantities of energy and water needs for the two scenarios are presented in Table 6.1. In Scenario 1, energy coverage, which is the first priority, reaches 79.16%, while water demand coverage is 62.95%. The total amount of energy covered throughout a year is 209,032 MWh, while water coverage reaches 2,655,388 m^3. In Scenario 2, energy coverage, being the last priority, is 72.58%, whereas water demand coverage, the main priority, reaches 99.99%. In this case, the total amount of energy covered is 191,657 MWh, while water coverage reaches 4,218,250 m^3.

TABLE 6.1

Coverage Rates and Quantities of Energy and Water Demand Under the Two Scenarios

	Scenario 1	Scenario 2
Energy coverage	79.16%	72.58%
Energy produced per year	209,032 MWh	191,657 MWh
Water coverage	62.95%	99.99%
Water produced per year	2,655,388 m^3	4,218,250 m^3

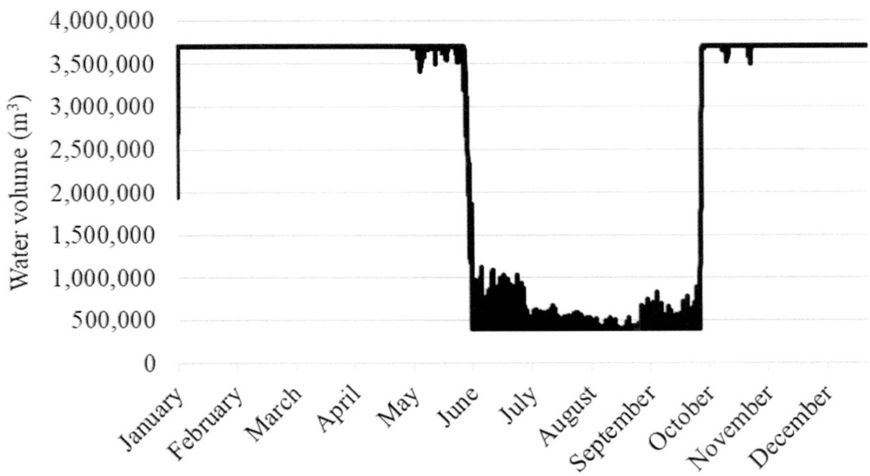

FIGURE 6.2
Annual variation of reservoir water volume under the first scenario (m^3).

Comparing the two (2) scenarios, it is observed that Scenario 2 achieves maximum water demand coverage (99.99%), since priority is given to water management. However, energy coverage is higher in Scenario 2 (79.16%).

In the first scenario, the variation of the reservoir water level is shown in Figure 6.2. The level remains high during the winter months but drops sharply from June onwards. This decrease coincides with the peak demand period, indicating that the stored water is not sufficient to cover the increased summer needs. The near depletion of the reservoir by October highlights the necessity for alternative water resource management strategies.

In the second scenario, the reservoir level shows greater stability compared to the first scenario, as shown in Figure 6.3. The total available water volume

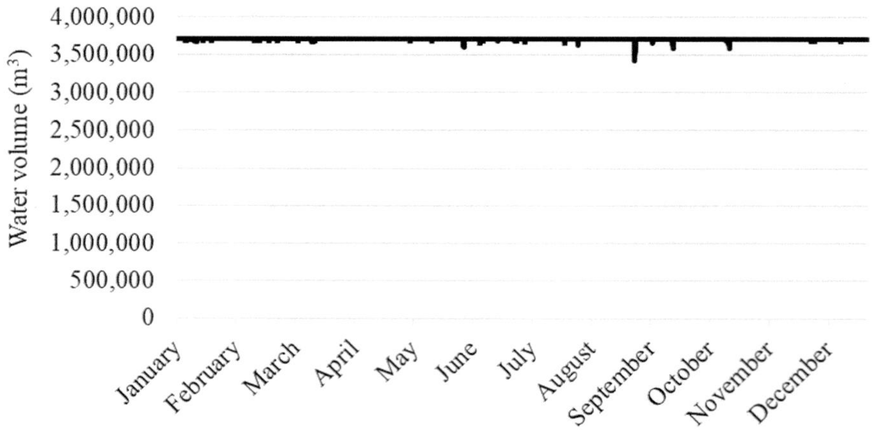

FIGURE 6.3
Annual variation of reservoir water volume under the second scenario (m^3).

remains almost constant at high levels throughout the year, without significant reductions, indicating more effective water resource management, as expected since priority is given to meeting water demand.

Between the two scenarios, significant differences are observed in the fluctuation of the reservoir level and, consequently, in the management of stored water. In Scenario 1, the reservoir level decreases sharply during the summer months due to increased water demand and hydropower production, reaching very low levels before being refilled toward the end of the year. In Scenario 2, the level remains high and stable throughout the year, indicating limited use of the reservoir either due to reduced energy production or restricted water withdrawals.

Figure 6.4 presents the hydroelectric energy produced by the turbine for each scenario, throughout the year. In Scenario 1, production shows strong fluctuations and increases significantly during the summer months, following the higher demand, reaching a peak of about 2,700 MWh in June. The total annual hydroelectric production in Scenario 1 amounts to 5,495 MWh. In Scenario 2, energy production is considerably lower compared to Scenario 1. Periods of high production are fewer, indicating that priority is given to preserving the water storage rather than generating electricity from the turbine. In Scenario 2, the turbine operates mainly to produce hydroelectricity for desalination, which explains the stability of the reservoir level. At the same time, some small production peaks appear during the summer months, when energy demand is higher, not exceeding 200 MWh, but these are limited in both duration and intensity. The total annual hydroelectric energy production in Scenario 2 is 194 MWh.

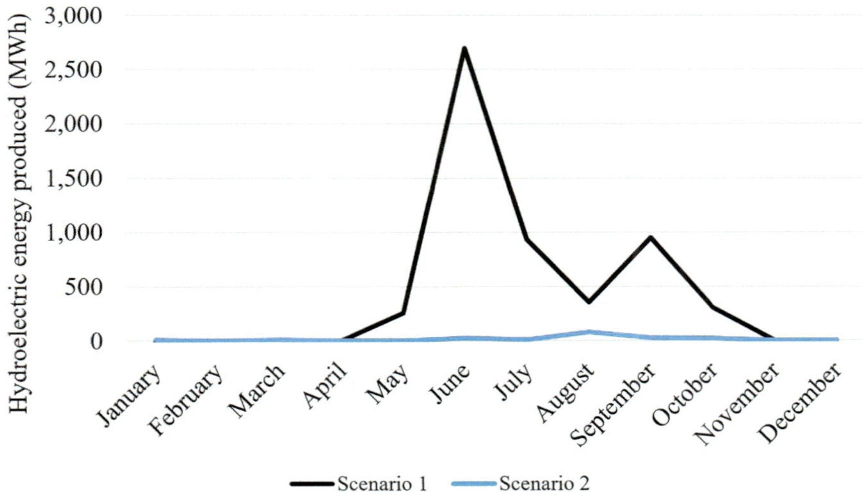

FIGURE 6.4
Hydroelectric energy production from the turbine under the two scenarios throughout the year (MWh).

6.4 Conclusions

The research work assessed the performance of "an HRES" combining wind power, desalination, and pumped-hydro storage on Mykonos, an island characterized by intense seasonal variability. Two operational scenarios were examined: one prioritizing electricity production (Scenario 1) and the other prioritizing water desalination (Scenario 2). The analysis revealed that the system's performance strongly depends on the selected management strategy, emphasizing the importance of aligning water and energy priorities to ensure resource security and enhance sustainability.

The comparison of the two scenarios shows clear variations between energy and water coverage. Scenario 2 fully secures water needs, achieving 99.99% coverage, as water management is prioritized. In contrast, Scenario 1 prioritizes electricity autonomy (79.16%) but limits water coverage to 62.95%, indicating insufficient availability under energy-first strategies.

The analysis of reservoir water levels highlights important differences among the two scenarios. In Scenario 1, the volume decreases sharply during the summer due to increased water demand and hydropower production, reaching very low levels before being restored toward the end of the year. In Scenario 2, the reservoir remains high and stable throughout the year, reflecting reduced water use and lower energy production as priority is given to water demand.

The comparison of hydroelectric production shows clear differences between the scenarios. In Scenario 1, output increases strongly during the summer, peaking at about 2,700 MWh in June, with total annual production of 5,495 MWh. In Scenario 2, production is minimal, limited mainly to supporting desalination, and reaches only 194 MWh/year. These patterns reflect the different management priorities, with Scenario 2 demonstrating the most active use of stored water to balance energy and water needs.

Since Mykonos is connected to the mainland grid through the Cyclades interconnection system, full energy independence is not required. Prioritizing water production, as in Scenario 2, represents a more practical and effective strategy, achieving almost complete water coverage, while maintaining sufficient energy performance.

Future research focuses on the application of climate change scenarios to assess how shifts in wind, solar, and water availability affect the performance of HRES. Comparative analysis of alternative storage technologies, including batteries and hydrogen, alongside pumped hydro provide further insights into long-term sustainability. Expanding the framework to different island regions with diverse geographic and resource characteristics enhances the applicability and generalizability of the results, supporting broader strategies for achieving energy and water autonomy.

References

Alexakis, D. E. 2022. Applying factor analysis and the CCME water quality index for assessing groundwater quality of an Aegean Island (Rhodes, Greece). *Geosciences*, 12(10), 384. https://doi.org/10.3390/geosciences12100384.

Babatunde, O., Denwigwe, I., Oyebode, O., Ighravwe, D., Ohiaeri, A., and D. Babatunde. 2022. Assessing the use of hybrid renewable energy system with battery storage for power generation in a University in Nigeria. *Environmental Science and Pollution Research*, 29(3), 4291–4310. https://doi.org/10.1007/s11356-021-15151-3.

Bertsiou, M. M., and E. Baltas. 2022. Management of energy and water resources by minimizing the rejected renewable energy. *Sustainable Energy Technologies and Assessments*, 52, 102002. https://doi.org/10.1016/j.seta.2022.102002.

Economou, A. 2010. Renewable energy resources and sustainable development in Mykonos (Greece). *Renewable and Sustainable Energy Reviews*, 14(5), 1496–1501. https://doi.org/10.1016/j.rser.2009.11.008.

Fulazzaky, M. A., Syafiuddin, A., Muda, K., Martin, A. Y., Yusop, Z., and N. H. A. Ghani. 2023. A review of the management of water resources in Malaysia facing climate change. *Environmental Science and Pollution Research*, 30(58), 121865–121880. https://doi.org/10.1007/s11356-023-30967-x.

Hellenic Statistical Authority – ELSTAT. 2021. https://www.statistics.gr/en/home/.

Kalogiannidis, S., Kalfas, D., Giannarakis, G., and M. Paschalidou. 2023. Integration of water resources management strategies in land use planning towards environmental conservation. *Sustainability*, 15(21), 15242. https://doi.org/10.3390/su152115242.

Kiehbadroudinezhad, M., Merabet, A., Hosseinzadeh-Bandbafha, H., and C. Ghenai. 2023. Environmental assessment of optimized renewable energy-based microgrids integrated desalination plant: considering human health, ecosystem quality, climate change, and resources. *Environmental Science and Pollution Research*, 30(11), 29888–29908. https://doi.org/10.1007/s11356-022-24051-z.

Lagaros, C., Papathanasiou, A. F., and E. Baltas. 2024. Implementing a hybrid renewable energy system to meet water and energy demand in Serifos Island. *European Water*, 87(88), 33–47. https://www.ewra.net/ew/pdf/EW_2024_87–88_04.pdf.

Mendez-Morales, M., Karipoglu, F., Ivanković, M., Lukić, T., and C. Rebelo. 2025. Transforming crete's sustainable energy landscape: a modular energy island approach. *International Journal of Energy Research*, 1, 2363366. https://doi.org/10.1155/er/2363366.

Papathanasiou, A. F., and E. Baltas. 2025. Achieving water and energy independence, economic sustainability, and CO2 reduction through hybrid renewable systems: a case study of Skyros Island. *Water*, 17(9), 1267. https://doi.org/10.3390/w17091267.

Papathanasiou, A. F., Bertsiou, M. M., and E. Baltas. 2024. Pumped-storage hydropower and hydrogen storage for meeting water and energy demand through a hybrid renewable energy system. *Euro-Mediterranean Journal for Environmental Integration*, 9(3), 1471–1483. https://doi.org/10.1007/s41207-024-00523-1.

Papathanasiou, A. F., Ntemiri, K., and E. Baltas. 2025a. Sustainable water resources management and energy production in Small Mediterranean Islands. *Environmental Processes*, 12(2), 1–18. https://doi.org/10.1007/s40710-025-00762-6.

Papathanasiou, A. F., Pasaniotis, G., and E. Baltas. 2025b. Comparison of hydro-pumped and green hydrogen as energy storage processes: a case study on Kefalonia Island, Greece. *Sustainable Chemistry and Pharmacy*, 48, 102216. https://doi.org/10.1016/j.scp.2025.102216.

Public Water Supply and Sewerage Company of Mykonos – DEYAM. 2025. https://deyam.gr/.

Qudah, A., Almerbati, A., and E. M. Mokheimer. 2025. Techno-economic analysis and optimization of standalone hybrid photovoltaic (PV)/wind turbine (WT) with water tank storage driven reverse osmosis (RO) desalination system. *Desalination*, 600, 118522. https://doi.org/10.1016/j.desal.2024.118522.

Skroufouta, S., Mavrogiannis, A., and E. Baltas. 2024. A methodological framework for the development of a hybrid renewable energy system with seawater pumped storage hydropower system under uncertainty in Karystos, Greece. *Sustainable Energy Technologies and Assessments*, 71, 104002. https://doi.org/10.1016/j.seta.2024.104002.

Tsiaras, E., Andreosatou, Z., Kouveli, A., Tampekis, S., and F. A. Coutelieris. 2025. Off-grid methodology for sustainable electricity in medium-sized settlements: The case of Nisyros Island. *Clean Technologies*, 7(1), 16. https://doi.org/10.3390/cleantechnol7010016.

Wulandari, R., Iswara, A. P., Qadafi, M., Prayogo, W., Astuti, R. D. P., Utami, R. R., Jayanti, M., Awfa, D., Suryawan, W. K., Fitria, L., and G. Andhikaputra. 2024. Water pollution and sanitation in Indonesia: a review on water quality, health and environmental impacts, management, and future challenges. *Environmental Science and Pollution Research*, 31(58), 65967–65992. https://doi.org/10.1007/s11356-024-35567-x.

7

Complementary Green Innovations for a Sustainable Future

Tadeusz Żaba and Jan W. Dobrowolski

7.1 Introduction

A sustainable future for humanity requires wider exchange of good practices related to efficient common actions of multidisciplinary teams of experts and knowledge-based society at local scale with international cooperation on protection of the biosphere focused on solving common problems contributing to sustainable management of the natural resources adopted to climate change. Wider exchange of good practices in application of complementary green innovations and circular bioeconomy could support this activity, which is useful to all human beings. Useful are experiences of long-term activity in this field, e.g., 57 years of problem-solving training in this field beginning with first National Summer School in this field in Poland in 1968 and 50 years of international training integrated with transdisciplinary case study in different regions initiated in 1973 at the first International Summer School in the Pieniny Mountains – the region of Europe's oldest border park, on the border between Poland and Slovakia (Dobrowolski 1971, 1972, 1976, 1982). The keynote introduction to this inter-generational mission was given by the worldwide precursor of the concept of sustainable development and transdisciplinary science *sozology* (following the Greek *sozeo* and *logos*), Prof. W. Goetel, Prof. J. Kostrzewski, who was at that time director of the World Health Organization (WHO) and later president of the Polish Academy of Sciences, and expert in environmental health and epidemiology. Thousands of interested university students and members of scientific non-governmental organizations (NGOs) supported by experts in more than 30 subjects of study took part in this voluntary education. A series of national and international schools and workshops were followed by many international conferences on sustainable development and common action at regional and global scales, focused on improvement of the quality of human life and environmental/nutritional health and dissemination know-how for early detection

DOI: 10.1201/9781003654889-7

risk factors, more efficient primary prevention, and creation of green jobs in different countries. The problem-solving training of experts was supplemented by lifelong learning for common action experts in natural, technical, social, and other sciences and knowledge-based sustainable society, e.g., for 25 years at the former AGH Open University, as well as by series of International Conferences in this field from 1989–2024.

More efficient prevention of anthropogenic contamination of water is necessary, both for sustainable management of water adopted to climate change, as well as for primary prevention of water-related health hazards. A good motivation for integration activity of experts from all complementary fields, local stakeholders, enterprises, self-governing and administrative bodies, environmental NGOs, etc., is the history of incurable diseases of civilization, e.g., connected with contamination of water and the human food chain by mercury poisoning (*Minamata*) and with cadmium poisoning (*itai-itai*).

Results of an etiology of Minamata disease, an effect of untreated wastewater containing mercury from a big chemical center and contamination for a long time of the Minamata Bay in Japan, were introduced by Japanese expert Dr. J. Ui at the first International School on Human Environment and Health in Poland in 1973, followed next year by "Open for All" Seminar at the Committee for Protection of Social Health at the Polish Academy of Science in Krakow with contribution of a team of Japanese experts. In 1975, after the 1st World Congress Scientists for Better Human Environment in Kyoto, J.W. Dobrowolski started cooperation with the leading expert on Minamata disease, Prof. M. Harada, to make a comparison of results of a transdisciplinary case study on contamination with mercury of the natural environment in selected regions (of more than 100,000 inhabitants) of the Kumamoto Prefecture and in South Poland, and the related health hazards (focused on prevention of congenital malformations in cooperation with the team headed by Prof. J. Niwelinski). In Poland, epidemiologist Prof. K. Janicki found a positive correlation between contamination of the natural environment in rural areas with mercury (as a component of an applied fungicide) and the rate of incidence of human leukemia, and Prof. J.W. Dobrowolski found a similar correlation with cattle leukemia. He also introduced – in cooperation with Prof. Harada – an innovative method of nutritional reduction of mercury risk hazard among inhabitants of the Minamata Region. A Japanese team of experts and victims of Minamata disease contributed in Poland to summer schools and to educational film *Living Silver*. This short film was awarded by the United Nations Environment Programme (UNEP). Prof. Harada was invited as a worldwide expert for consultative visits in rivers and lakes in areas contaminated in Brazil and Canada. WHO recommended prevention of contamination with mercury as a global priority in 2018. In 2020, after 45 years of Japanese-Polish cooperation on innovative methods for prevention of Minamata disease, the team of Japanese scientists, local experts, and representatives of victims of Minamata disease contributed to open seminars in the city of Krakow in Poland.

The Krakow Water enterprise for efficient prevention of waterborne diseases integrated a highly efficient wastewater treatment system with online monitoring of drinking water quality (including, e.g., control of trace metals). There are good practices in very efficient cooperation among Japanese and Polish experts in complementary fields of modernized monitoring of environmental contamination, as well as in better treatment of wastewater. A result of this cooperation was introduced to Poland in 1981 as the first country in Europe introduced by Prof. Amaya and Prof. Sugiura to Japanese methods of screening monitoring of some air and water pollutants harmful to human health, as well as to nature and cultural heritage. These methods were valuable tools for coordinated by Prof. Dobrowolski's 33-year study including personal exposition and related health hazards, as well as the influence of traffic output on the inclusion of Krakow on the United Nations Educational, Scientific and Cultural Organization (UNESCO) list of World Heritage sites, along with selected cultural heritage sites in old cities in England, Italy, Greece, Spain, etc. Working with Prof. Amaya Eng. Tanuma, president of Bando Company of wastewater treatment plants, in Yokohama in Japan which introduced to Europe very innovative, efficient, economic and low energy consuming biotechnology for underground treatment of wastewater and prevention of related ecological and health hazards. Very efficient was also the long-term cooperation of scientists from Grece and Poland focused on interdisciplinary studies on complementary eco-innovations for a sustainable future in complementary aspects of bio-inspirated activity on local and global scale, including bioethics, biopolicy, and bioeconomy, in cooperation with founding president of the International University for the Bio-Environment, Prof. A. Vlavianos-Arvanitis (Dobrowolski 1991), better environmental and nutritional health focused on biomedical function of elements in cooperation with Prof. S. Ermiodou-Pollet and Dr. S. Pollet (Dobrowolski 2001b) and model reclamation of the oldest in Europe mining area of high cultural value in cooperation with Prof. C. Cassios (Dobrowolski 2001c, 2008; Belsare et al. 2023). This is human-oriented international mission of export as good practice to supplement by similar experiences and to follow for the future sustainability.

7.2 Methods and Analysis

7.2.1 Introduction of Innovation in Wastewater Treatment, Water Management, and Renewable Sources of Energy

Water supply and sewage disposal is one of the main elements of the proper functioning of a city. The inhabitants of Krakow use about 170,000 m^3 (Report of Kraków Water 2022) of water every day and produce about 220,000 m^3 of sewage. Water is treated in four treatment plants, and the daily production capacity is over 280,000 m^3. The total volume of wastewater captured in the municipal sewage system is treated in two central wastewater treatment plants and five local treatment plants. The Płaszów Sewage Treatment Plant is one of the largest in Poland. Its biological part of the treatment plant has a capacity of 328,000 m^3 per day, and in the rainless season about 165,000 m^3 of sewage flows through it per day, with the maximum flows of 656,000 m^3 per day during heavy rainfall. The technological part of the plant is located on an area of 50 ha.

Wastewater is sent through the sewage system to the wastewater treatment plant, where it undergoes a treatment process that consists of two main technological processes: the mechanical and biological treatment of wastewater. After a complicated technological process, the treated wastewater is directed to the receiver, and the residue of the treatment process is municipal sewage sludge. On average, about 260 tons of dewatered sludge is generated daily at Krakow's wastewater treatment plants to a content of about 24% dry matter. In terms of dry matter, this is about 64 tons. About 20,000 tons of dry sludge matter remain to be managed per year. The applicable regulations do not allow for its storage, and the maximum period of its stay at the place of production cannot be longer than one year. Therefore, the sludge requires further development. The best solution for its management is thermal disposal while maintaining all environmental protection requirements. In order for the sludge to be thermally disposed of, it must meet certain conditions. The dry matter content of the dewatered sludge must be between 22% and 26%. At the Thermal Sludge Utilization Station in Krakow, the sludge is subjected to the process of thermal utilization using the fluidization phenomenon. The advantages of this process are the even distribution of the entire stream of supplied air over the cross-sectional surface of the reactor, good mixing conditions in the sand layer, and ensuring optimal contact between the burned waste and the supplied air. During the process, heat is generated, which is used to heat the fluidization air introduced into the air chamber and to generate steam for the steam dryer used to pre-dry the sludge to a dry matter content of approximately 34%. Waste heat, on the other hand, is used to heat technological facilities in the treatment plant.

Kraków Waterworks is a modern and leading company among the biggest water enterprises in Poland. It is financially stable, uses state-of-the-art technologies, and carries out multimillion-zloty investments as part of long term strategies. Nowadays, the company stands as a synonym for quality and safety of supplied services. Respect toward the environment's resources and its responsible use are values transmitted to younger generations by the Kraków Waterworks.

The company publicizes the high quality of its products and the safety of Kraków's residents in terms of water supply through the "In Kraków good water straight from the tap" campaign. The campaign "It won't fit in a sewer!" highlights the issues of the use of urban sewerage and sewage treatment. Every year, the Kraków Waterworks is a winner of prestigious competitions and plebiscites. For many years, the company has been awarded with the "Fair Play Enterprise" title. Innovative approach to reducing energy consumption and wider use of renewable energy sources integrated with sustainable water management is a priority in the activities of the Krakow Water. The model activity focuses on sustainable water supply and wastewater collection as well as the approach to pro-environmental activities.

Electricity is an important element of the costs of water supply and sewage collection. This is particularly important in recent times, as energy prices are rising sharply. This makes it necessary to constantly look for savings. One of the possibilities is the use of innovative solutions. These include both those that allow for reduced energy consumption and those that allow its production. With this in mind, new technologies promoting renewable energy sources are systematically introduced – e.g. energy of gravitational potency of treated wastewater and modern cogeneration systems operating on the basis of biogas generated in the wastewater treatment process – and thanks to the use of modern solutions in control systems, energy efficiency is increased by reducing electricity consumption.

Krakow Water is involved also in research-developing studies for reduction both operating cost and use of energy also by modern solutions in the field of energy efficiency and the use of energy generated in the process of waste management as innovative contribution to circular economy and sustainable development.

Facilities where the Krakow Water save or generate electricity include the following.

1. Experimental photovoltaic farm
2. First stage pumping station
3. Second stage pumping station
4. Blower station
5. Water turbine
6. Cogeneration
7. Biogas turbine
8. Thermoelectric generator
9. Stations for thermal utilization of sewage sludge

An experimental photovoltaic farm has been operating in the largest sewage treatment plant in Krakow for 11 years, thanks to which we have full knowledge about the effectiveness of this type of solutions in our climatic conditions. The photovoltaic farm allows for an annual production of about 68 MWh of electricity and assessment of the possibility overturning expenditures on this type of investment.

Another element that allows for reduction of energy consumption and carbon dioxide (CO_2) emissions is the integrated energy efficiency system (IEES) used to control the technological devices consuming the most electricity, which by analyzing the efficiency of individual devices on an ongoing basis controls their operation so that at the moment those with the highest efficiency work. The operation of the first stage pumping station is managed by the IEES, which – based on the collected measurement data – analyzes the efficiency of a particular pump and decides which of the pumps will be switched on at a given moment. Thanks to this, we can use less electricity to pump out the same volume of water. The same system also manages the work of the second stage. The pump with the highest efficiency works all the time, while periodically, according to technological needs, the system attaches another pump. The blower control system is also an important element of the system which supplies the necessary amount of the air to the oxygen chambers of biological reactors.

The control algorithm allows the pressure set point to be changed depending on the oxygen demand of the bioreactor chambers. In addition, the so-called power-on queuing has been implemented. The blower with higher efficiency, which is calculated in the system, is set as the first to be added in the queue.

When calculating the blower efficiency, the following factors are taken into account: diffuser opening degree, current power flow, pressure difference on the filter, and energy coefficient kW/m^3. Additionally, in order to prevent the possibility of overloading the low-voltage switching station, a protection has been introduced that prevents switching on of three drives supplied from one section of the R21NN switchgear. The control system will not allow the blower to start if two drives in a given section are already running.

The next system generating electricity in the treatment plant is a water turbine located at the outlet of treated sewage. It is a Kaplan turbine with double regulation by a steering wheel with movable blades and a rotor with a variable blade position, coupled with a generator with a power of 85 kW. In the rainless season, the average power generated by the turbine is approx. 60 kW, and in the first year of operation, the turbine produced 380 MWh of electricity.

The main element producing renewable energy is a cogeneration system, which consists of two cogeneration units with a rated electric power of 800 kW and a thermal power of 810 kW, and two biogas turbines with a capacity of 65 kW each. The task of the turbines is to use excess biogas and eliminate its burning in the candle. In total, this system generates 4,300 MWh of electricity and provides heat for all sewage treatment plant facilities. Thanks to the production of energy in renewable sources, the waterworks of the city of Krakow reduced CO_2 emissions by 2,257 tons, sulfur dioxide (SO_2) by 2.11 tons, carbon monoxide (CO) by 0.6 tons, and dust emissions in total by 0.127 tons.

7.3 Results of Case Study

7.3.1 Results of the Introduction of Green Innovation in the Krakow Waterworks as a Model Study

An experimental photovoltaic farm has been operating in the largest sewage treatment plant in Krakow for 11 years, thanks to which we have full knowledge about the effectiveness of this type of solutions in our climatic

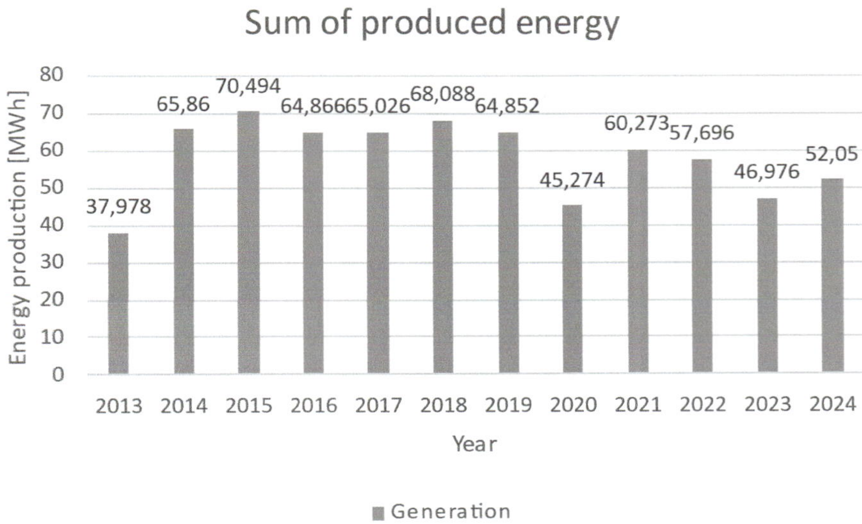

FIGURE 7.1
Electricity production during the years 2013–2024 in an experimental photovoltaic farm.

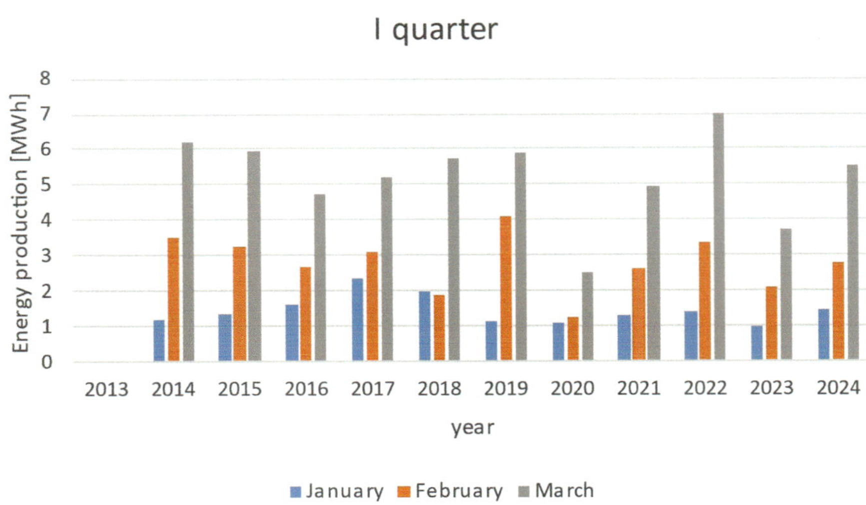

FIGURE 7.2
Electricity production during the first quarter of 2013–2024.

conditions. The photovoltaic farm allows for an annual production of about 68 MWh of electricity and assessment of the possibility of returning expenditures on this type of investment. Figure 7.1 shows electricity production in the years 2013–2024 in an experimental photovoltaic farm.

On the other hand, Figure 7.2 presents a comparison of the amount of electricity generated in individual quarters of 2013–2024.

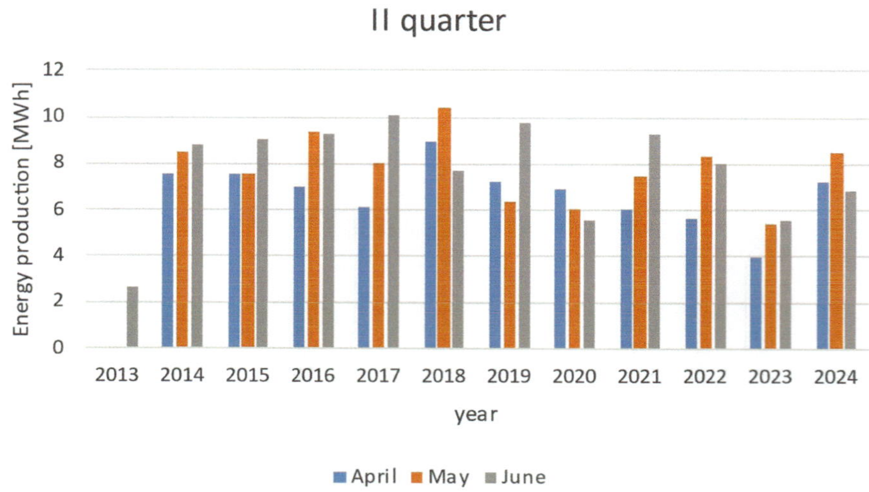

FIGURE 7.3
Electricity production during the second quarter of 2013–2024.

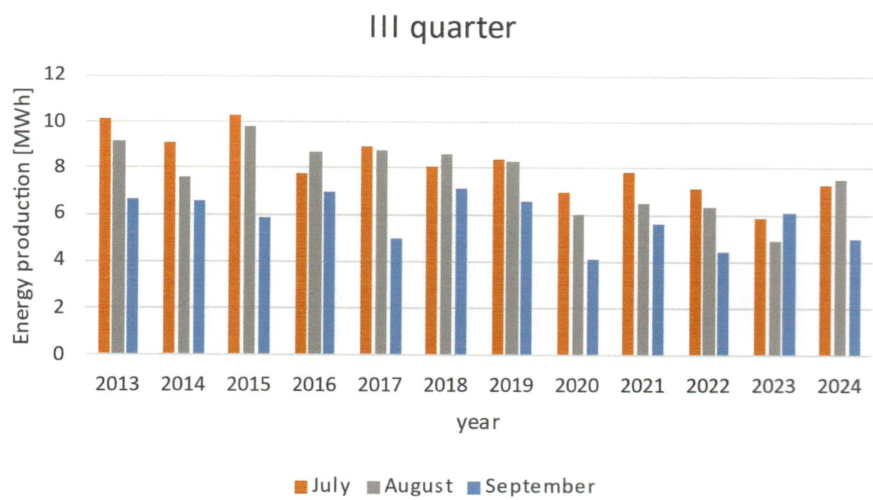

FIGURE 7.4
Electricity production in the third quarter of 2013–2024.

We can compare the production in first quarters to the energy produced in second quitter of the previously mentioned period. Data for the second quarter are represented in Figure 7.3.

Figure 7.4 Presents electricity generated during the third quarter. We can clearly see that the trend of increase in production starting from April is kept through summer months.

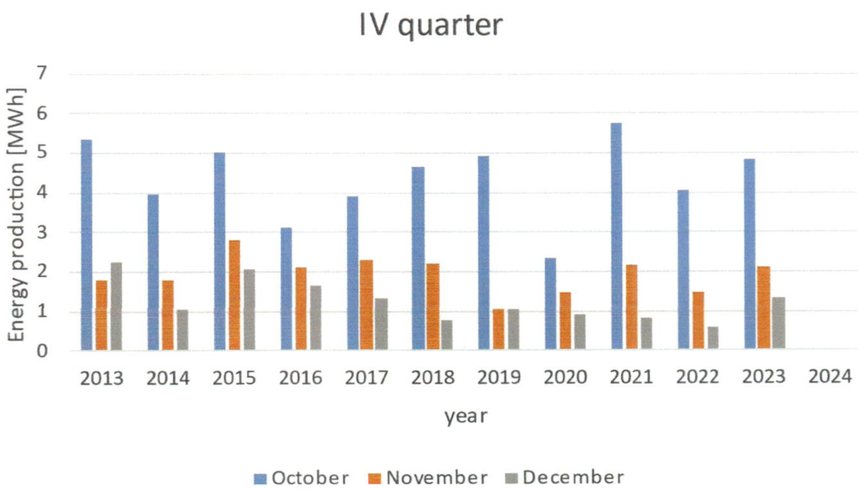

FIGURE 7.5
Electricity production during the fourth quarter of 2013–2024.

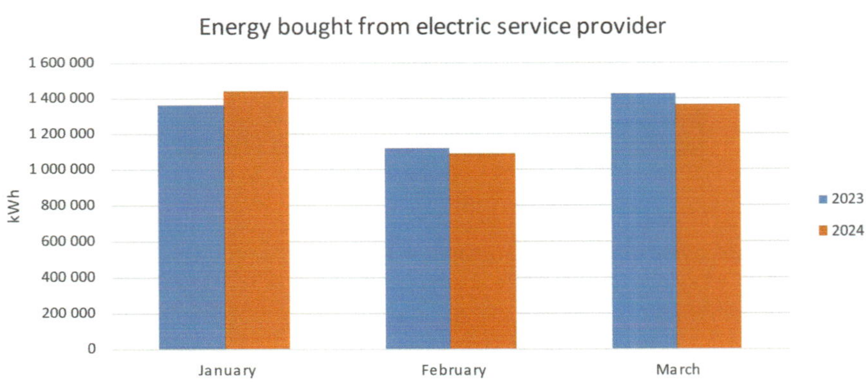

FIGURE 7.6
Comparison of the amount of electricity purchased from the supplier during the first quarters of 2023 and 2024.

In Figure 7.5 – which represents generation in October, November, and December – we can see decrease in production during late autumn and winter.

In the first quarter of 2024, two more CHP units with a capacity of 800 kW each were commissioned. Figure 7.6 shows the impact of the commissioning of new CHP units on the purchase of electricity from the supplier.

Figure 7.7 shows the amount of green electricity produced by individual generating units in the first quarter of 2024 (JK1–JK 4 CHP energy produced

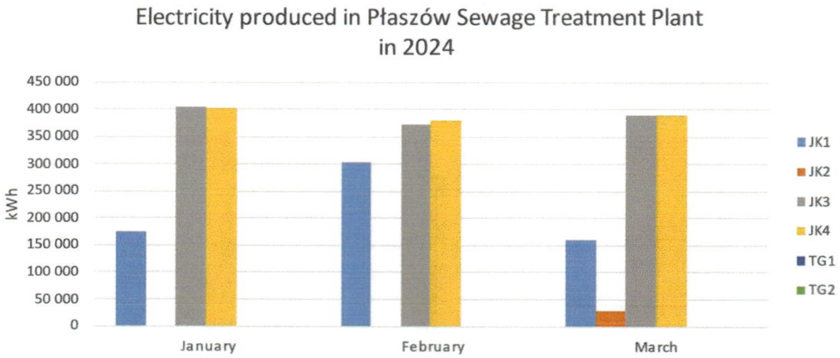

FIGURE 7.7
Production of green electricity by individual generating units at the Płaszów wastewater treatment plant.

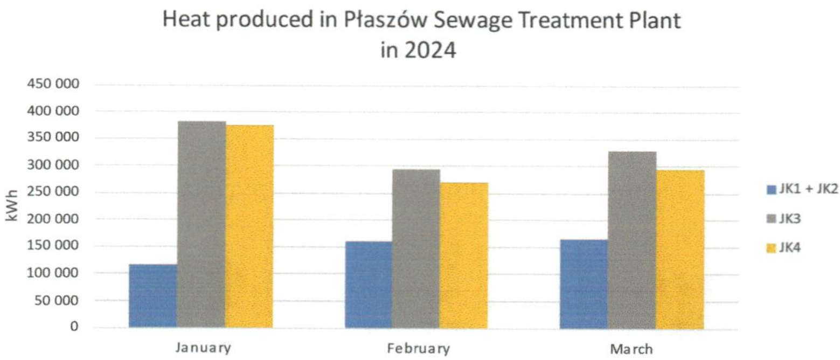

FIGURE 7.8
Production of green heat during the first quarter of 2024 by cogeneration units operating at the treatment plant.

by units with a capacity of 800 kW, TG1–TG2 energy produced by biogas-powered turbines).

Figure 7.8 shows the production of green heat by individual CHP units.

It should be emphasized that *all energy generated from renewable sources* – both electricity and heat – is used for the technological process, thus reducing purchases from suppliers. In terms of thermal energy, the treatment plant is 100% self-sufficient.

7.4 Discussion

A key factor for wide-scale application of complementary innovations for common action of experts and knowledge – based on a sustainable society focused on sustainable management of natural resources and better health and quality of life – seems to be *lifelong learning* following 57 good years of practice. The crucial innovation in both professional training and open widely available education is the replacement of traditional task-oriented learning with problem-solving learning integrated with transdisciplinary case study. The output of these studies contributed both to progress of knowledge and efficiency of common action for benefit to all.

Case study is an innovative good practice promoting modernization of sustainable water management integrated with promotion of renewable sources of energy with linkage to training of experts on national and international scales, as well as its being open for the inhabitants of all age groups education supporting the promotion of green innovation.

Wide-scale use of laser biotechnology may be useful for reclamation including mangrove phytocenosis protecting coastal regions, as well as for acceleration growth rate and formation of protective "green screens" consisting of maidenhair *Ginkgo biloba* for protection against large-scale transfer of fair destruction common during dry periods in many countries, including Greece. Laser biotechnology was adopted by Prof. J.W. Dobrowolski and his students for worldwide promotion of sustainable development of model rural areas in Brazil, China, Cuba, Egypt, France, Germany, Georgia, Greece, India, Japan, Kazakhstan, Laos, Madagascar, Malaysia, Mexico, Myanmar, Nepal, Norway, Peru, Poland, Syria, Turkey, Uzbekistan, and other countries. Such *eco-innovation is especially useful for prevention against both desertification and flooding, as well as for creation during a short time of many new green jobs and promotion of the bio-based circular bioeconomy*, in particular for developing countries all over the world (e.g., in regions of high unemployment, semiarid areas) – including conceptual, model international centers of bioeconomy and sustainable development adopted to climate change in selected cities and rural regions, including reserves of nature in India (in cooperation with Prof. D.K. Belsare) and in semiarid regions of Brazil (in cooperation with Prof. J.O.B. Carioca).

7.5 Conclusions

According to more than 50 years of international training exchange of complementary green innovations and lifelong problem-solving learning by experts, a knowledge-based society is a key factor for common action for promotion of a sustainable future – both on regional and international scales.

A case study of an integrated low-energy system based on cogeneration and renewable sources of energy (for reduction of emissions of greenhouse gases) and sustainable management of good quality water at the Krakow Waterworks is integrated with education on local and international scales. The novelty of this study is that it is related to introduction of an empirically selected algorithm of low-power laser biostimulation for improvement of both bioremediation of wastewater and enhancement assimilation of carbon dioxide, and for "greening" and sustainable bio-management regions of lakes, rivers, and areas out of use as green innovation increasing retention of water and adaptation to climate change.

Integration of the application of laser biotechnology with complementary green innovations could be useful for development of common action on sustainable management of water and the biological resources, for dissemination of good practice among regions facing similar problems, and for improvement of nutritional health (including more efficient primary prevention of incurable diseases related to contamination of water) and for quality of life based on innovative circular bioeconomy–driven sustainable development.

7.6 Recommendations

Promotion of a good quality of green innovation is recommended for development of the digital economy, including high-level development on an inter-regional scale (Bai et al. 2023).

Worldwide implementation of green innovations may to be a key factor for adopting recommendations of the United Nations Secretary-General for the internationally agreed-upon target of 1.5°C reduction in temperatures; according to the experts, 22 billion tons of CO_2 must be reduced from the currently projected total in 2030 (UN Secretary General 2023).

Better adaptation to climate change related to increasing risk of extreme conditions – occurrences of drought or great flooding – requires improvement of dissemination and application of complementary know-how on complementary green technologies focused on adequate international online training and integration of sustainable management of water and other natural resources for better environmental and nutritional health, including recommendations of human ecology (Dobrowolski 1972, 1982; Fukushima et al. 1976; Dobrowolski et al. 1976), ecotoxicology (Coulston and Korte 1976; Dobrowolski and Fertig 1990; Dobrowolski and Rozanowski 1998), environmental biotechnology (Dobrowolski 2001c, 2001d, 2007b; Dobrowolski et al. 2004, 2012a, 2012b; Tursunov and Dobrowolski 2015), and circular bioeconomy (Belsare et al. 2023), as well as ecological engineering (Dobrowolski 2000, 2001c, 2001d; Dobrowolski and Zielinska-Loek 2002; Dobrowolski et al. 2017a) and *medical elementology* (Dobrowolski 1997, 2001b; Vohora and Dobrowolski 1990).

Better adaptation to such environmental factors has to be oriented on biomimetics – both related to new biomaterials as well as to innovative sustainable design following, e.g., protective external structures following materials and construction of exoskeletons for some animals (invertebrates and vertebrates) well adapted to very strong external powers in water (e.g., in coastal regions) and on land ecosystems existing under extreme conditions (e.g., more resistant to earthquakes, very strong winds, etc.).

Development of interdisciplinary and international cooperation on new transdisciplinary *biomimetic biotechnology* (circular bioeconomy–driven sustainable development integrated with human ecology and many biological and chemical disciplines, and also integrated with psychological, sociological, and economic study and supported by proper management) would be a key factor for success of such human-oriented missions (integrating eco-innovative achievements of experts in material engineering, architecture, geoengineering, computer science, information technology (IT), and many complementary research fields) contributing to optimization of greening cities, including urban agriculture.

A very useful design for a sustainable future for big cities would be innovative underground centers of circular bioeconomy–driven sustainable development. Such centers should integrate highly efficient bio-treatment wastewater and organic waste into habitats, with production also in the city of pollutant-free food such as vegetables grown using hydro-botanic cultivation in underground greenhouses using renewable sources of energy such as biogas as from outputs of sewage and waste bio-treatments.

We intend to introduce a working hypothesis such that a starting point for solving these crucial problems (for many regions of the world) would include the following.

1. Research-developing study of improvement on a macro scale in vitro culture of stem cells involved in process of synthesis of proper biomaterials, including bio-nanotechnology.
2. Construction of model shelters (including life-supporting systems adopted to climate change and future living houses, habitats, etc.),
3. Optimization of sustainable design of innovative construction as a result of cooperation among experts in architecture, mathematical modeling, and IT supported by social and natural sciences, etc.
4. Test study of new generation buildings resistant to strong earthquakes and wind for heuristic improvement of innovative constructions and their function.
5. Supplementing innovative life-supporting systems as a starting point for a new generation of green habitat in the model area (a little similar to circular environmental biotechnology for wastewater treatment, cleaning, and regeneration of the air for reuse, bio-management of waste, etc. recommended for long-term manned space missions), adopted to conditions of the model region of study.

We have to collect complementary achievements in biotechnology-based wastewater treatment and reuse, as well as wastes bio-management for biofuel and bioenergy production useful for protection of optimal climatic conditions in indoor environments of tested buildings and also for underground greenhouses for fixation of CO_2 and production of pollutant-free vegetables, mushrooms, algae, and aquaculture for breeding fish, etc., as well as for the introduction of evergreen underground gardens for healthy recreation and integration of eco-habitats.

A proposed solution would be also useful all over the world for better adaptation to climate change based on protection of homeostasis based on negative bio-feedback systems, as well as for improvement of environmental/nutritional health and for sustainable labor markets related to circular bioeconomy–driven sustainable development, including sustainable use of water and other adaptations to climate change.

One of the key factors for a sustainable future is wide-scale application of complementary green innovations in sustainable design (Dobrowolski and Kobylarczyk 2013) based on modern environmental monitoring, transdisciplinary human ecology, and environmental biotechnology (Dobrowolski 2017), including new perspectives of integrated sub-molecular biomonitoring related to mechanisms of interaction between natural and anthropogenic electromagnetic fields and living organisms (Popp and Zhang 2000), as well as ecosystems and ecological balance.

Contemporary tendencies in architectural design are minimizing impact on the natural environment (including protection of ecological balance and biodiversity) and reduction of use of natural resources and energy (with given preference to renewable sources of energy). The crucial problems in sustainable design are optimization of conditions in housing environments and surroundings for human life and environmental health (Kobylarczyk 2018a). Environmental conditioning in the design of residential areas is integrated with greening cities with linkage to common action for better quality of life and integration of sustainable society (Dobrowolski and Kobylarczyk 2013).

Better adaptation of cultivated plants and soil microorganisms to extreme environmental conditions connected with climate change requires additional energy. Empirically selected algorithms of low-power laser photostimulation (of high energy density) is a new cheap and very useful tool for enhancement of food production, which is good for nutritional health, as well as production of biomass for bioenergy production under suboptimal conditions including water deficiency. It is also better for reclamation of deteriorated areas out of use (including the introduction of energy plantations in mining and in industrial regions and on salted soil), for increased water phytoremediation and for enhancement of assimilation of greenhouse CO_2, for more efficient bioremediation and biodegradation of pollutants in contaminated soil (including much more efficient biodegradation of carcinogenic PAH in soil and water) and better reclamation of areas contaminated by petrochemical pollutants. It is better as well for more efficient bioremediation of toxic metals, and also for wastewater biotreatment (e.g., by microalgae, local water plants based on hydrobotanic treatment, and reuse of water, e.g., for irrigation of agricultural areas or aquaculture, as well as for improvement environmental health and better protection of biodiversity).

Over the years advanced computational simulations esp. backed by High Performance Systems (HPC) were used in multiple use-cases such as medical, technological and environmental, such as processing of data provided by European network of satellites for remote sensing [38–43].

7.6.1 Proposals for Worldwide Application

The mentioned long-term experiences and a new concept of underground urban agriculture could be a starting point for internet-based long distance training of experts (e.g., online dissemination of know-how) and for foundation of an international network of teams of experts and innovative enterprises for wide-scale applications of laser biotechnology and complementary eco-innovations for sustainable adaptation of local biological resources to climate change (in linkage with the circular bioeconomy) and for improvement of quality of life of millions of people – including populations under risk of malnutrition; nutritional deficiency of necessary trace elements such as iron, zinc, selenium, etc.; as well as those facing risk of more common environmental-borne diseases (e.g., incurable Minamata disease and common infectious diseases such as iron-deficiency anemia) and hunger in many regions of the world.

As a result of transdisciplinary case studies initiated and coordinated by the authors, a new application of laser biotechnology was introduced for a statistically significant increased amount of deficient, biologically active trace elements in different species of cultivated plants such as vegetables, fruits, maize, flax, potatoes, etc. This new tool is very useful for nutritional prevention of some common diseases connected with deficiency of such elements like iron, zinc, selenium, etc. We found also that in result of application different algorithms of laser irradiation, bioaccumulation of toxic trace metals by irradiated plants could be reduced. Practical outputs of both discoveries could be useful for adaptation to different side effects of climate change and for nutritional prevention of immunosuppression and the risk of infective diseases pandemics [36].

The results of our experiments on different applications of environmental and health friendly laser biotechnology are useful for stimulation of germinating seeds, and growth rates of seedlings and cuttings of different plants, as well as for better symbiosis with a consortium of soil bacteria and fungi, as an innovative eco-engineering contribution to suboptimal physico-chemical conditions connected with anthropocentric deterioration of the natural environment (including enhancement resistance to water deficiency and to different pollutants of the air, soil, and water).

Our long-term experimental study with laser biotechnology for sustainable development may be also useful for acceleration reforestation and biodiversity protection, as well as for more efficient greening cities and their regions in linkage with better adaptation to climate change.

Laser biotechnology is reasonable in cost/benefit analysis and also very useful for enhancement of both production of pollutant-free feed for animals (under sub-optimal conditions) and for stimulation reproduction of different animals. A result of application of empirically selected algorithms of low-power laser irradiation is efficient stimulation of ovary function and ovulation in different species of animals. Prof. Dobrowolski and team headed by the top expert in artificial insemination, Prof. Laszczka at the National Institute of Zootechnics in Krakow, Poland, discovered very significant reduction of mortality of spermatides of some breeding mammals (after standard cryoconservation using liquid nitrogen) and stimulation of mobility of spermatides as a result of application of a proper algorithm of irradiation by selected laser diodes and mobile medical lasers. Complementary laser biostimulation could be widely used for stimulation of reproduction of cows and other breeding animals (including better artificial insemination), as an innovative contribution of pollutant-free food production and reduction malnutrition.

These results could be also useful for more efficient reproduction of some rare animals under risk of extinction in linkage to ecological principles of protection of biodiversity.

Our experiments also opened a new perspective for enhanced reproduction of different species of invertebrates – cultivated on organic wastes – both as more and more popular elements of human diet and also for feeding fish in aquaculture (in linkage with laser biotechnology for better bio-treatment of sewage, as well as for stimulation growth and biomass production by microalgae).

Integration of application laser biotechnology in complementary fields could be supplemented by system approach to introduction also eco-innovative know-how in aquaculture, as well as in apiculture, with linkage to cultivated medical plants and introduction of some bee products (including new herbs and honey rich in essential trace elements). Common actions of experts from different fields and nations, as well as local knowledge-based society, could contribute in creative ways to sustainable management of the biological resources, dissemination of good practice among regions facing similar problems and for improvement of nutritional health and quality of life for all based on the innovative circular bioeconomy.

7.6.2 Funding

This research did not receive any financial support from university, financial agencies in the public, commercial, or non-for-profit sectors.

7.6.3 Credit Authorship Contribution Statement

Jan W. Dobrowolski: Introduction, lifelong learning towards sustainable future, water-borne health hazards, innovations for sustainable development and reduction environmental health hazard.

Tadeusz Żaba: Case study on green innovations for sustainable water management.

The Krakow Waterworks PLC as subject for green innovation, methods of introduction innovation in wastewater treatment, water management and renewable sources of energy, results of introduction green innovation in the Krakow Waterworks as model study, visualization.

Both authors: Conclusions.

7.6.4 Declaration of Competing Interest

The authors declare that they have no known competing financial interest or personal relationship that could have appeared to influence the work reported in this chapter.

References

Bai W., Hou J., Xu J., Chen J., Does digital economy development successfully drive the quality of green innovation in China? *Polish Journal of Environmental Studies*, 32 (3), 2001–2014, 2023.

Belsare D., Dobrowolski J.W., Saini S., eds., *Circular Bioeconomy and Sustainable Development*, Nagar, Delhi, 2023.

Bubak M., Kasztelnik M., Malawski M., Meizner J., Nowakowski P., Varma S., Evaluation of cloud providers for VPH applications. *In: 13th IEEE/ACM International Symposium on Cluster, Cloud and Grid Computing*, pp. 200–201, IEEE, 2013.

Coulston F., Korte F., eds., *Environmental Quality and Safety*, Academic Press, Stuttgard, 1976.

Dobrowolski J.W., 20 lat działalności Uniwersytetu Otwartego AGH jako próba edukacji wszystkich dorosłych grup wiekowych (20 years of activity of the AGH Open University as trial of education of all age groups of adults) (in Polish). *In: Edukacja dorosłych jako czynnik rozwoju społecznego*, T. Aleksander ed., Vol. 1, pp. 201–218, Jagiellonian University, Krakow, 2010.

Dobrowolski J.W., Action of the Polish university youth on man and environment (in Polish). *Biuletyn Polskiego Komitetu do Spraw UNESCO* (*Polish Committee for UNESCO Bulletin*), 7, 1971.

Dobrowolski J.W., Application of laser biotechnology in environmental management. *In: Biotechnology. The World Congress on Biotechnology*. vol. 3, Section VI Environmental Biotechnology, Society for Chemical Engineering and Biotechnology, Berlin, 2000.

Dobrowolski J.W., The conception of the International University for the bioenvironment. *In: Biopolitics and Bioenvironment*, A. Vlavianos-Arvanitis, ed., Vol. III, pp. 326–331, BIO, Athens, 1991.
Dobrowolski J.W., Ecotoxicology, human ecology, laser biotechnology in primary prevention of environmental health hazard. *Przeglad Lekarski*, 58 (Sup.7), 1–4, 2001a.
Dobrowolski J.W., Ecotoxicology-medical elementology-laser biotechnology and primary prevention of environmental health hazard. *In: Proceedings of the 3rd Symposium on Trace Elements in Humans: New Perspectives*, S. Ermidou-Pollet, S. Pollet, eds., pp. 135–143, University of Athens, 2001b.
Dobrowolski J.W., Environmentally-friendly biotechnology as a key to modernization of environmental engineering. *Inżynieria Środowiskat*, 6 (z.2), 259–272, 2001c.
Dobrowolski J.W., Perspectives of application of laser biotechnology in management of the natural environment. *Polish Journal of Environmental Studies*, 10 (Sup.1), 7–9, 2001d.
Dobrowolski J.W., Environmental monitoring, human ecology as well as biotechnology as the premises for pro-health development of the environment. *In: Urban Housing Environment*, G. Schneider-Skalska, E. Kusinska, eds., pp. 111–120, Cracow University of Technology, Krakow, 2017.
Dobrowolski J.W., Environmental risk factors of cancer and their primary prevention. *In: Health and Toxicology, Advances in Environmental Control Technology Section*, P.N. Cheremisinoff, N.P. Cheremisinoff, eds., Chapter 1, pp. 1–46, Gulf Pub Company, Houston, 1997.
Dobrowolski J.W., For the future. *Poland*, 2, 3–5, 1972.
Dobrowolski J.W., *The Importance of General Sozological Education for Modernization Education of University Students* (*Concept of the Interuniversity Center of Sozological Education*) (in Polish), pp. 53–57, *The Importance of General Sozological Education for Modern Education of University Students in Linkage to UNESCO Recommendations*, pp. 121–122, *The Importance of Ecological Education for Ecological Consciousness of Society. Five Years of Polski Klub Ekologiczny, 1980–1985*, Akademia Gorniczo-Hutnicza im. S. Staszica, Krakow, 1985.
Dobrowolski J.W., The influence of laser photostimulation of plants of bioaccumulation of elements. *In: Proceedings of the 7th International Symposium "New Perspectives in the Research of Hardly Known Trace Elements"*, I. Pais, ed., pp. 47–52, University of Horticulture and Food Industry, Budapest, 1996.
Dobrowolski J.W., *Innovative Biological Monitoring and Laser Biotechnology for Sustainable Development of Rivers Regions*, La Sapienza University, Roma, 2006.
Dobrowolski J.W., An interdisciplinary study of, and education for, the sustainable development of national park regions in Poland and a new concept of sound tourism management applied to the Cinque Terre National Park in Italy. *In: Across Disciplinary Bounders Towards a Sustainable Development*, E.A. Stuhler, O' Suilleabhain, D.J. De Tombe, S. Mistra, eds., pp. 89–108, Reiner Hampp Verlag, Munchen, Meri, 2007a.
Dobrowolski J.W., Perspective of the application of innovative environmental biotechnology for sustainable development of co-operating regions. *Geomatics and Environmental Engineering*, 1 (4), 77–88, 2007b.
Dobrowolski J.W., Polish experiences on interdisciplinary cooperation among scientists and regions for promotion sustainable development by promotion innovative environmental technologies (English Summary) Polskie doświadczenia wzakresie interdyscyplinarnej współpracy naukowców z regionami dla

promocji zrównoważonego rozwoju a perspektywy zastosowania innowacyjnych technologii środowiskowych. *In: Ochrona i Inżynieria Środowiska-Zrównoważony Rozwój (Environmental Protection Engineering and Sustainable Development)*, Vol. 35, pp. 7–19, Monografie Wydziału Inżynierii Mechanicznej i Robotyki AGH, 2007c.

Dobrowolski J.W., Modern science for future environmental and social activity of university youth. *In: Environmental Quality and Safety, Global Aspects of Chemistry, Toxicology and Technology as Applied to Environment*, F. Coulston, F. Korte, eds., Vol. 5, p. 78, Georg Theme Pub, Stuttgart, Academic Press, New York, San Francisco, London, 1976.

Dobrowolski J.W., The Polish initiatives. *Naturopa*, 61 (9), 1989.

Dobrowolski J.W., Reflexions sur la qualite de Vie et la perennite de la Civilisation Contemporaine. *In: L'apres Developpement Durable Espaces, Nature, Culture et Qualite*, A. Da Lage, J.-P. Amant et al., eds., pp. 167–180, Ellipses, Paris, 2008.

Dobrowolski J.W., Thinking about the future. *Poland*, 2, 1982.

Dobrowolski J.W., Bedla D., Czech T., Gambus F., Górecka K., Kiszczak W., Kuzniar T., Lewicki P., Mazur R., Sliwka M., Tursunov O., Wagner A., Wieczorek J., Zabochnicka-Swiatek M., Integrated innovative biotechnology for optimization environmental bioprocesses and green economy. *In: Integrative Biology Book Series, Optimization in Environmental Bioprocesses*, H.J. Purohit, et al., eds., Springer, Berlin, 2017a.

Dobrowolski J.W., Kobylarczyk J., Wagner A., Mazur R., Involving diverse stakeholders for sustainable development; some learning experiences from areas across Poland. *In: Optimizing Open and Distance Learning in Higher Education Institutions*, U.C. Pandey et al., eds., IGI Global, Hershey, 2017b.

Dobrowolski J.W., Budak A., Bogusz B., Evaluation of the influence of laser light on selected pathogenic moulds in vitro (in Polish with English summary). *Inzynieria Srodowiska*, 4 (1), 115–124, 1999.

Dobrowolski J.W., Budak A., Trojanowska D., Rymarczyk M., Macuda J., Laser stimulation of *Trichophyton mentagrophytes* for the enhancement biodegradation of hydrocarbons. *Journal Environmental Engineering and Management*, 11 (10), 1783–1788, 2012a.

Dobrowolski J.W., Sliwka M., Mazur R., Laser biotechnology for more efficient bioremediation, protection of aquatic ecosystems and reclamation of contaminated areas. *Journal of Chemical Technology & Biotechnology*, 87, 1354–1359, 2012b.

Dobrowolski J.W., Fertig S., Bioindication of the contamination by heavy metals of people and animals in the regions of some metallurgical works, Polish academy of science-Krakow section. *Proceedings of Mining and Geodety Commission, Geodety*, 35, 112–118, 1990.

Dobrowolski J.W., Glowiak B., Michajlow W., Human environment scientific congress-the first complex scientific congress on environmental problems (in Polish). *Nauka Polska*, 5, 1976.

Dobrowolski J.W., Guha A.S., Open University and modern distance learning in Poland and India for sustainable development. *Geomatic and Environmental Engineering*, 18, 26–35, 2011.

Dobrowolski J.W., Kobylarczyk J., Sustainable design, modern environmental protection engineering and biotechnology. *Technical Transactions*, 12, 2013.

Dobrowolski J.W., Rozanowski B., *The Influence of Laser Light on Accumulation of Selected Macro-Trace and ultra Elements by Some Plants. Menege und Spurenelemente*, pp. 147–156, Friedrich-Schiller-Universitat, Jena, 1998.

Dobrowolski J.W., Rozanowski B., Zielinska-Loek A., The influence of low power laser light of some trace elements in plants. *Biological Bulletin of Poznan*, 33, 1996a.

Dobrowolski J.W., Wachalewski T., Smyk B., Barabasz W., Różycki J., Experiments on the influence of laser light on some biological elements of natural environment. *Environmental Management and Health*, 8 (4), 136–141, 1996b.

Dobrowolski J.W., Rozanowski B., Zielinska-Loek A., Sliwka M., Gowin K., Mazur R., Lewicki P., Zakrzewska A., Slazak A., Perspectives of application of laser biostimulation for more bioremediation of soil and wastewater. *In: International Conference on Bioremediation of Soil and Groundwater*, pp. 133–148, Politechnika Slaska, Krakow, 2004.

Dobrowolski J.W., Smyk B., Environmental risk factors of cancer and their primary prevention. *Journal of Environmental Pathology, Toxicology and Oncology*, 12 (1), 55–57, 1993.

Dobrowolski J.W., Wolkowski Z.W., Żaba T., Primary prevention of new pandemic and biomimetic-based adaptation to situation connected with COVID-19 pandemic. *Journal of Environmental and Life Sciences*, 9 (6), 133–140, 2020.

Dobrowolski J.W., Zielinska-Loek A., *The Laser Photostimulation of Willow Cuttings Planted Alongside Main Roads and Change of Concentration of Elements in the Willow's Organs. Mengen und Spurenelemente*, pp. 21, 334–340, Friedrich-Schiller-Universitat, Leipzig, 2002.

Fukushima Y., et al., eds., *Scientists for Better Human Environment*, Science Council of Japan HESC, Tokyo, 1976.

Kobylarczyk J., *Environmental Conditioning in the Design of Living District*, Cracow University of Technology, Krakow, 2018a.

Kobylarczyk J., *Environmental Conditions in the Design of Residential Area*, Cracow University of Technology, Krakow, 2018b.

Malawski M., Meizner J., Bubak M., Gepner P., Component approach to computational applications on clouds. *Procedia Computer Science*, 4, 432–441, Elsevier, 2011.

Meizner J., Bubak M., Malawski M., Nowakowski P., Secure storage and processing of confidential data on public cloud. *In: Parallel Processing and Applied Mathematics*, pp. 272–282, Springer, Berlin, Heidelberg, 2014.

Meizner J., Malawski M., Bubak M., Flexible and secure access to computing clusters. *Computer Science*, 11, 21–36, Akademia Górniczo-Hutnicza im, Stanisława Staszica w Krakowie, University of Science and Technology Wydawnictwo AGH, 2010.

Meizner J., Malawski M., Ciepiela E., Kasztelnik M., Harężlak D., Nowakowski P., Król D., Gubała T., Funika W., Bubak M., Mikołajczyk T., Płaszczak P., Wilk K., Assel M., Virolab security and virtual organization infrastructure. *In: International Workshop on Advanced Parallel Processing Technologies*, pp. 230–245, Springer, Berlin, Heidelberg, 2009.

Meizner J., Nabożny M., Radecki M., Szepieniec T., Zdybał M., *Recent Advances of the Cloud Platform Delivered in the Infrastructure as a Service Model for the PL-Grid Scientific Communities, e Science on Distributed Computing Infrastructure: Achievements of PLGrid Plus Domain-Specific Services and Tools*, pp. 54–60, Springer International Publishing, Berlin, 2015.

Popp F.A., Zhang J.Z., Mechanism of interaction between electromagnetic fields and living organisms. *Science in China, Series C Life Science*, 5, 2000.

Tursunov O., Dobrowolski J., Addressing environmental issue and risks in Uzbekistan. *Integrated Journal of Engineering Research and Technology*, 2 (1), 62–69, 2015.

UN Secretary General, 20 November 2023. http://www.un.org/sg/en/content/sg/press.encounter/2023-11-20/secretary-general-press-conferenceunep-emissions-gap-report-launch.

Vohora S.B., Dobrowolski J.W., eds., *New Horizon of Elements and Health*, Hamdard University Press, New Delhi, 1990.

8

Open Innovation for Circular Economy Transition

Christos Kalantzis, Konstantinos G. Aravossis, and Christos Nikoloudis

8.1 Introduction

The transition to a *circular economy (CE)* is recognized as a pathway for achieving sustainable development and a green transition, within the framework of the European Green Deal. However, this transition is far from straightforward. It requires systemic change research, cross-sectoral collaboration, and new innovative business models that go beyond firm-centric or closed research and development (R&D) paradigms. In this context, *open innovation (OI)* emerges as a promising approach that can catalyze the co-creation of circular solutions, enable knowledge exchange across organizational and sectoral boundaries, and overcome the fragmentation of innovation systems.

OI – defined broadly as the purposive inflow and outflow of knowledge to accelerate innovation processes – has gained traction in both academic discourse and industrial practice. When applied to CE transitions, OI can support the design of closed-loop systems, promote industrial symbiosis, and foster business model innovation that separates economic growth from resource consumption.

This chapter aims to explore how OI can actively support the transition to CE at multiple implementation levels: *macro* (policy and governance), *meso* (industrial networks and ecosystems), and *micro* (organizational strategies and operations). Drawing on insights from a literature review, it identifies key trends, barriers, and future research directions, offering insights for academics, practitioners, and policymakers involved in sustainable innovation and development.

 DOI: 10.1201/9781003654889-8

8.2 Conceptual Foundations

8.2.1 Circular Economy: Principles and Implementation Levels

It is widely accepted that CE can tackle climate change and other global challenges – like biodiversity loss, waste management issues, and pollution – by decoupling economic activity from the consumption of natural resources. In a CE, materials are kept in circulation for a longer time than in a linear economy, thus aiming to minimize the use of virgin material and the generation of waste (Harris et al., 2018).

CE is a concept with no scientific background, as it was not introduced through a scientific experiment or methodology. More than 100 definitions can be found in the literature, which are not opposing, but complementary conceptually (Kirchherr et al., 2017). Various research and institutional initiatives across the world have delved into CE, publishing their own insights and definitions on the subject.

The Ellen MacArthur Foundation is an Institute with a considerable research impact on CE and its implementation practices, which describes it today, as a system whereby materials never become waste and nature is regenerated. In a CE, products and materials are kept in circulation through processes like maintenance, reuse, refurbishment, remanufacture, recycling, and composting (Ellen MacArthur Foundation, n.d.). The Institute bases CE on the following three principles, driven by design.

- Eliminate waste and pollution
- Circulate products and materials
- Regenerate nature

Considering that CE is an economic model, its implementation needs to reach every economic activity simultaneously for a successful transition. In terms of research and institutionally speaking, CE implementation occurs at three distinct but interrelated levels, as shown in Figure 8.1 (Ahmadov et al., 2025; Dautremont and Gobbo, 2025; Dolgova and Nikitaeva, 2025; Drofenik et al., 2025).

- **Macro level:** The highest level of implementation that includes policy instruments, legislation, and regional/national strategies that shape an enabling environment for circular transition.
- **Meso level:** The implementation level that includes collaboration across firms, sectoral innovation platforms, and supply chain restructuring.
- **Micro level:** The lowest level of implementation that includes practices in the organization level, product redesign, reverse logistics, and circular business models. The Micro level also includes the consumer's behavior.

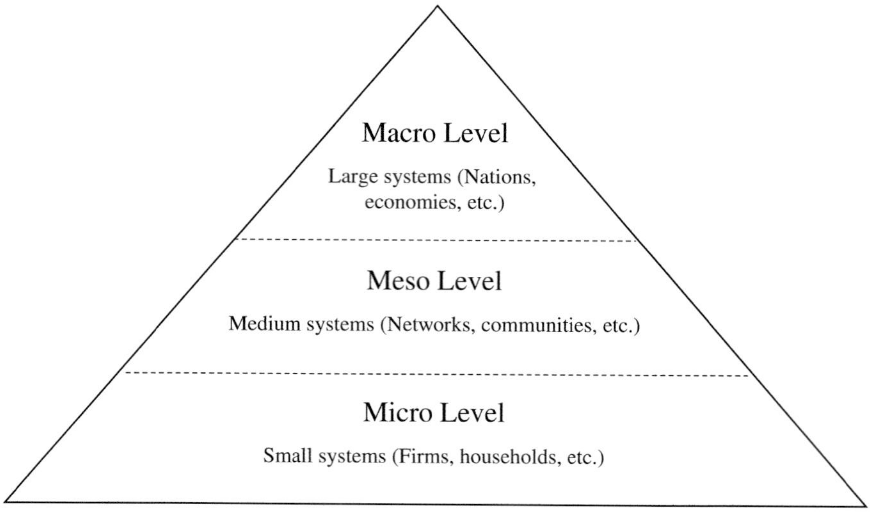

FIGURE 8.1
Policy implementation levels.

Source: Author's illustration

8.2.2 Open Innovation: From Theory to Practice

Originally conceptualized by Henry Chesbrough, OI challenges the traditional view that valuable innovation must originate from within an organization. Instead, it encourages openness to external knowledge sources and collaborative mechanisms (Jesus and Jugend, 2023), including the following.

- **Inbound OI:** Leveraging external ideas and technologies for the organization's purpose.
- **Outbound OI:** Externalizing internal knowledge and IP for societal impact or profit.
- **Coupled OI:** Co-creation through partnerships, alliances, and innovation ecosystems. Combines the mechanisms and characteristics of both inbound and outbound OI.

This categorization, as also shown in Figure 8.2, is widely used in literature to describe the innovation and knowledge flows during OI processes.

When applied to sustainability and CE, OI expands its scope from organizational performance to environmental and social value creation, introducing concepts such as mission-oriented innovation, quadruple helix extroversion strategy, and open sustainability innovation.

Although collaboration may require additional time and resources, it is widely recognized as a critical element in the circular transition, as

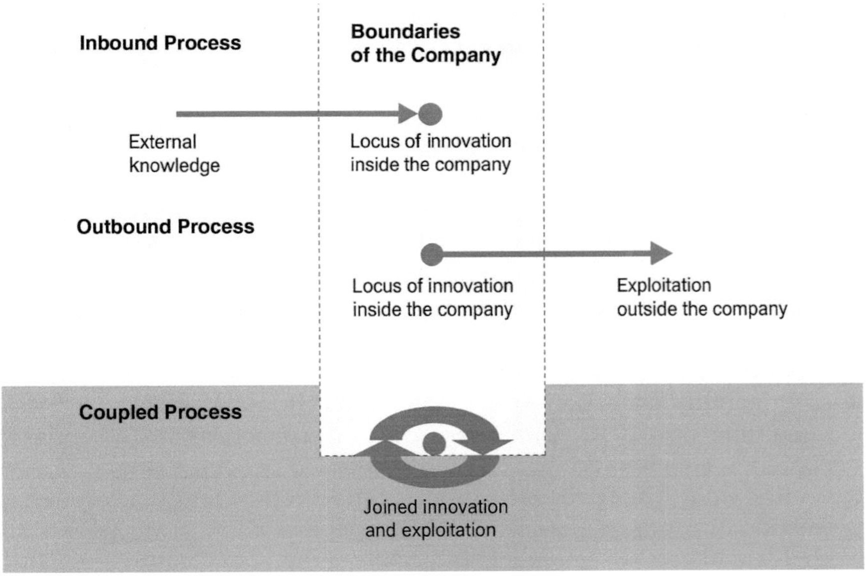

FIGURE 8.2
Classification of open innovation practices based on knowledge flows.

Source: Nerone et al. (2014)

collaboration enables more effective outcomes. Collective work is defined as a system in which stakeholders are aware that they are working as a group to achieve a common contribution, and OI can enable the needed collective work toward a sustainable economy (Jesus and Jugend, 2023).

8.3 Open Innovation for Circular Economy at the Macro Level

At the macro level, decision-makers in regions, nations, and international organizations play a critical role via policy and legislation frameworks to create the conditions for a just and effective CE transition. OI can support this transition by facilitating stakeholder engagement towards participatory policymaking and public-private collaboration, ensuring that all available knowledge and insights from the stakeholders are utilized. Key mechanisms include the following.

- *Living labs* and *policy co-design platforms*, as citizens, businesses, and governments jointly make and evaluate policy guidelines.
- *Open science and data ecosystems* that enable transparency, benchmarking, and evidence-based policymaking.

- *Regulatory sandboxes* and regions that allow for experimentation with circular business models and technologies in a controlled environment.
- *Monitoring initiatives* at a high level to monitor and evaluate the results of policies and act accordingly.

Living labs and collaboration platforms are not only experimental environments, but also democratic innovation hubs for the co-creation of policies. In the CE context, they enable high-level decision-makers to anchor strategies in real-life societal and economic needs, as well as industry and market needs. Evidence from existing research shows that such collaborative ecosystems improve both the legitimacy and acceptance of policies, which is crucial for the sustainability transition of systems (Almirall and Wareham, 2008). At the same time, open science and open data infrastructures are essential for transparency, benchmarking, and evidence-based policymaking, grounded in scientific work. The European Commission's Strategy for Data emphasizes the creation of common European data spaces that allow governments and stakeholders to access comparable indicators (European Commission, 2020). For CE, particularly, this includes monitoring resource efficiency, waste generation, or material flows, enabling fair comparison across countries and regions. A monitoring framework provides a mechanism for governments to learn from outcomes, adjust policies dynamically, and ensure transparency and accountability (Kirchherr et al., 2017).

Regulatory sandboxes provide safe conditions to test suggested practices and innovative technologies, as well as business models, before they are included in policy. In the context of CE, such controlled environments allow experimentation with new forms of reuse, sharing, or material recovery, helping policymakers detect potential technological, societal, or legal barriers and adapt regulations in ways that remain innovation-friendly. Several EU member states are already using sandbox approaches for green technologies, turning them into policy laboratories for circular transitions (European Environment Agency, 2021).

Governments and regional authorities can use OI not only to consult stakeholders but also to co-create legislation, ensuring that regulatory frameworks remain adaptive and innovation-friendly. The literature explores current national strategies affecting circular transitions and distinguishes *five critical areas* that act as policy action themes too at the macro level: water, energy, emissions, waste, and materials. These areas, in fact, categorize actions at meso and micro levels, too, according to current practices (Drofenik et al., 2025). Monitoring initiatives are framed based on these critical areas, especially in the European Union, including member countries and relevant institutions.

Although the implementation of policies and legislation has been researched to this day, no case studies can yet be found in the literature. Current publications focus only on strategic or conceptual frameworks, on how to utilize OI on the national level and large systems to foster systematic,

TABLE 8.1

OI Mechanisms at the Macro Level and Expected CE Outcomes

Main OI Mechanisms	Circularly Expected Outcomes
Living labs and policy co-design platforms	Participatory governance and multi-stakeholder engagement
Open science and open data ecosystems	Transparency, benchmarking, and evidence-based policymaking
Regulatory sandboxes	Safe experimentation with new circular technologies and business models
Monitoring frameworks and European data spaces	Adaptive, innovation-friendly, and accountable CE regulation

sustainable change, including the transition towards a CE. Table 8.1 presents the OI mechanisms mentioned in the literature that contribute to the implementation of CE.

8.4 Open Innovation for Circular Economy at the Meso Level

At the meso level, circularity emerges through inter-organization collaboration and sectoral transformation. In this case, networks of the related stakeholders at the sector level are engaged in the transition. Here, OI acts as a catalyst for the following.

- **Industrial symbiosis:** Sharing of resources, by-products, and energy among organizations within an industrial cluster.
- **Innovation networks:** Formation of consortia, hubs, and collaborative platforms such that companies and research organizations can co-develop new circular solutions and practices.
- **Circular supply chains:** Digital infrastructures to use blockchain, the Internet of Things (IoT), and data-sharing platforms to trace material flows and enable transparency, as well as to introduce known sustainable practices.

The main meso-level OI mechanisms and their corresponding circular outcomes are summarized and presented in Table 8.2.

The concept of industrial symbiosis is most common in factory-level applications, but can also be applied to smaller scales with small and medium-sized enterprises (SMEs). This practice can transform industrial or non-industrial clusters into circular ecosystems whereby waste from one economic activity becomes input for another. The Kalundborg model in Denmark – featuring

TABLE 8.2

OI Mechanisms at the Meso Level and Expected CE Outcomes

Main OI Mechanisms	Circularly Expected Outcomes
Industrial symbiosis and upcycling networks	Resource and energy efficiency, by-product reuse, waste reduction, regenerative practices.
Innovation networks and consortia	Shared circular practices, systemic innovation
Blockchain and IoT-enabled supply chains	Trust, traceability, and transparent material flows
Cross-sector collaborative platforms	Lower transaction costs, stronger inter-firm collaboration

by-product exchanges across energy, pharmaceutical, and agricultural firms – illustrates how OI can optimize resource efficiency and reduce waste (Haskins, 2006).

At the same time, there is a steady global trend towards a sustainable transition via an eco-innovation ecosystem, reinforced by Industry 4.0 technologies. Innovation networks extend beyond being simple collaboration networks to become inclusive ecosystems where multiple stakeholders co-develop circular solutions. Empirical research shows, in fact, that meso-level innovation ecosystems – empowered by digital technologies – enable deeper integration and systemic innovation across firms and sectors (Surya et al., 2025).

In most cases, a deep and systemic integration of the CE at the meso level requires sustainable supply chains. In literature, digital platforms like blockchain and IoT are known for supporting circular supply chains, as they enable immutable tracking of material flows, boost transparency, and facilitate shared data ecosystems (Mankata et al., 2025). For example, a blockchain-enabled model in the construction industry enhances trust and traceability, while broader systematic reviews show blockchain's potential to operationalize the 4R principles (reframe, redesign, realign, and reinforce) in business models, offering practical change.

Challenges at this level include competition issues, aligning incentives across sectoral value chains, and developing shared metrics for circular performance for monitoring and evaluation procedures. Nevertheless, successful examples from eco-industrial parks and cross-sector alliances demonstrate that OI principles can significantly lower the transaction costs of collaboration.

In an empirical study, researchers collected data from 13 organizations, organized in two different networks, that worked on creating and establishing upcycling practices for food and beverage waste. Two interorganizational collaborative approaches have been examined for enabling regenerative systems, while highlighting the roles of coordinators and orchestrators in the collaborative processes. The case study concluded that OI mechanisms are crucial in fostering circular business development and networks (Perotti et al., 2025).

TABLE 8.3

OI Mechanisms at the Micro Level and Expected CE Outcomes

Main OI Mechanisms	Circularly Expected Outcomes
Open organizational culture	Alignment around CE goals, openness to external knowledge
Transformational leadership and participatory governance	Co-created circular solutions, empowered employees
Circular business model innovation	New circular value chains and sustainable business models
Crowdsourcing and innovation challenges	Expanded problem-solving capacity, faster adoption of circular practices
Consumer co-design and OI programs	Increased consumer acceptance, engagement in CE solutions

8.5 Open Innovation for Circular Economy at the Micro Level

At the micro level, the implementation of OI practices within individual organizations is crucial for embedding CE principles into day-to-day operations. Key elements include the following.

- **Organizational culture:** A culture of openness, experimentation, and collaboration is essential for integrating external knowledge and co-developing circular solutions.
- **Leadership and governance:** Transformational leadership styles and inclusive decision-making structures foster internal alignment around CE goals.
- **Business model innovation:** OI enables the development of new circular business models (e.g., sharing platforms, remanufacturing services) through collaboration with customers, suppliers, and research institutions.
- **Consumer behavior:** A consumer, being a person, a household or an organization, plays a vital role in shaping the economy by affecting the demand–supply principle.

An overview of the main micro-level OI mechanisms and their corresponding circular outcomes are highlighted in Table 8.3.

A culture of openness and intensified knowledge flows is a prerequisite in all organizations for developing and implementing circular solutions at the micro level, given how dynamic the global economy is. Innovative organizational cultures emphasize trust, collaboration, and risk tolerance as critical factors for organizations to integrate external knowledge and co-develop circular solutions within a network (Martins and Terblanche, 2003).

Towards an open culture, transformational leadership and governance are essential for aligning organizations around sustainability and circular economic goals. Since the beginning of the century, it has been known that environmental leaders adopt participatory and inclusive decision-making practices, enabling organizations to integrate sustainability into governance structures. Thus, leadership and governance via open and sustainable values actively foster a shared vision that empowers employees to contribute to OI initiatives and implement circular principles into daily operations (Egri and Herman, 2000).

Of course, having a circular business model is also critical as the base of a sustainable economic activity at both micro and meso levels. OI mechanisms such as crowdsourcing and partnerships can support firms in experimenting with new business models, or uptaking knowledge from existing circular ones. In the literature, all circular business models rely on collaboration across organizational boundaries, with some indicative archetypes being the product-service system, remanufacturing services, and sharing platforms (Lüdeke-Freund et al., 2019).

Finally, a CE demand at the market needs to be shaped by the consumers. It is often highlighted that consumer acceptance and participation are essential conditions for scaling circular practices (Geissdoerfer et al., 2017). At the micro level, OI initiatives allow firms to tap directly into consumer insights and co-create sustainable solutions.

Organizations adopting OI for CE implementation may use mechanisms such as crowdsourcing, innovation challenges, or strategic alliances to expand their problem-solving capacity and accelerate the implementation of circular practices at the micro level.

Batik SMEs are culturally and economically significant in Indonesia, but face sustainability challenges. The organizational performance of 70 batik SMEs in Banyuwangi, Indonesia, was researched to investigate the effect of OI and CE practices combined. The study highlighted insightful results for both research and practice, indicating that both OI and CE applications have a positive impact on the performance of companies. In fact, for those SMEs, particularly, the most valuable benefits were the resource optimization and innovation networks (Rahmat et al., 2024).

8.6 Challenges and Barriers to Open Innovation in Circular Transitions

Despite its potential and successfully implemented cases, the integration of CE practices via OI is not without obstacles. Common challenges in literature include the following

- **Knowledge and capability gaps:** There are challenges observed in integrating external technologies with existing knowledge and processes (absorptive capacity), as well as limited internal knowledge cultivation capabilities of SMEs (innovation capacity) (Jesus and Jugend, 2023; De Jesus and Mendonça, 2018; Yoshino et al., 2025).
- **Cultural and behavioral barriers:** Negative influence of existing culture on collaborations on innovation adoption, often due to lack of trust or risk-free choices (Geissdoerfer et al., 2017; Skordoulis et al., 2020; Bocken and Ritala, 2022).
- **Structural and market inertia:** Market and circular supply chain not developed yet, to offer safe investment conditions for commercialization (Triguero et al., 2022; Chistov et al., 2021; Bogers et al., 2020).
- **Governance, legal, and intellectual property (IP) issues:** IP management issues and licensing difficulties (Eisenreich et al., 2021; Brown et al., 2021; Patala et al., 2020).
- **Coordination and stakeholder misalignment:** Complexity in aligning goals and incentives among diverse stakeholders, due to different goals and technological maturity (Brown et al., 2020; Perotti et al., 2025; Köhler et al., 2022).
- **Technological and infrastructural limitations:** Technological barriers, such as a lack of appropriate technology and equipment, by the organizations (Antikainen et al., 2017; Suchek et al., 2021; Zhang et al., 2021).
- **External conditions and systemic constraints:** External socio-political and economic conditions and uncertainty, as well as a lack of environmental information on new practices (De Jesus and Mendonça, 2018; Geerken et al., 2019; Jakobsen et al., 2021).

Due to insufficient environmental knowledge and skills among employees and adopters, there are gaps in knowledge and capabilities that stand as obstacles to the integration of external solutions. Also, a gap in literature and practice is observed: a lack of theoretical frameworks, as well as practical implementation. Existing complexity in operationalizing CE models within and cultural inertia in adopting collaborative models are notable challenges, too, especially in cases of firms where digitalization is unachieved. On a high level, there is difficulty in integrating economic sustainability (life cycle costing, or LCC) with environmental sustainability (life cycle assessment, or LCA).

The influence of existing negative culture – due to prior lack of success – on external collaboration is often found as a barrier. Those cultural barriers often come from the lack of trust in the altruistic motives of business or anti-industry bias, and the risk of opportunistic behavior due to openness. Potential loss of competitive advantage from open knowledge sharing shapes the reluctance of small organizations, while dominant sectors are reluctant as well to change and adopt eco-innovations and circularity. The "open innovation paradox" of balancing knowledge sharing with protection is an issue that all stakeholders confront. Social dilemmas and confusion in innovation commons, as well as misalignment between firms and community needs, are barriers to changing the market.

Having a CE market and supply chain not developed, the high costs and resource requirements for innovations and commercialization are becoming high risks for investors and market shapers. Research highlights a disconnection between circular models and real-world applications, mainly due to the absence of large-scale demonstration projects. Lack of established frameworks, uncertainties about financial benefits, and overcoming the self-referential nature of architecture and design are some of the other recognized challenges.

IP management issues and a lack of proper production licensing pose a challenge to translating open technologies into open governance, discouraging the stakeholders from engaging in OI practices. At the same time, there is complexity in aligning goals and incentives among diverse stakeholders who are not on the same level of maturity and technological incompatibility, introducing stakeholder involvement and collaboration issues.

Technological barriers – such as a lack of appropriate technology and equipment – are quite often mentioned in literature, as well as the lack of technical solutions. At the same time, external socio-political and economic conditions and uncertainty make it more difficult for organizations or systems to foster change.

Among the most common barriers are the missing enabling policies, regulations, and financing needed to support incentives of CE through OI practices, as well as a variety of stakeholder collaboration issues, such as goal misalignment. These obstacles often directly counter the enablers promoted by OI. For instance, while stakeholder collaboration is a central tenet of OI, misalignment and lack of trust were identified as recurring issues. Similarly, although digital tools and platforms are praised for enhancing circular innovation, technological limitations and a lack of digital maturity in considerable cases remain barriers in many sectors.

Despite this, the literature points to promising prospects. OI – especially in the form of ecosystems, co-creation, and collaborative networks – can help bridge knowledge gaps, improve absorptive capacity, and accelerate the diffusion of sustainable practices. Frameworks like communities of practice (CoP) can foster regional innovation and policy alignment, which is crucial for systemic CE adoption – yet, realizing these prospects requires addressing structural, cultural, and policy-related barriers in parallel. In short, OI offers a powerful pathway for CE implementation – but its effectiveness depends on resolving the very barriers that currently inhibit circular transformation.

Overcoming these barriers requires deliberate strategies, including the following.

- Capacity-building and training programs focused on OI and sustainability.
- Clear governance frameworks for data sharing and IP protection.
- Incentives and funding mechanisms that reward collaboration for CE outcomes.

8.7 Future Research Directions

Considering the number of barriers to successfully implementing CE via OI, there are several key areas where further research is needed, as the literature suggests. Interdisciplinary and practice-oriented research is particularly valuable to bridge the gap between theory and implementation.

Regarding the *macro level*, research should explore how to align regulatory conditions among regions and countries, while also ensuring a consistent monitoring and evaluation framework. Moreover, forming open data and science ecosystems that support evidence-based policymaking has plenty of room for research, while more studies are needed to determine and design adequate funding tools and frameworks to support a just transition. Finally, achieving systemic resilience to global challenges using OI and CE principles is a promising subject for research.

Research on the *meso level* of implementation should support multi-firm collaborations by both aligning their goals, culture, and innovation maturity, as well as investigating the field of by-products reuse, to create more industrial symbiosis opportunities. Circular supply chain research can be further developed by integrating new digital methods and circular models, focusing on feasibility and adoption issues.

Considering the **micro level** and the absorptive capacity of organizations to new innovations and knowledge, research should focus on interventions such as learning mechanisms, as SMEs often lack the internal skills to adopt external resources. On the same page, more work is needed on participatory governance and transformational leadership, considering the impact of these two variables on the organization's culture. Another important research topic for scholars is balancing openness with IP protection, so that firms will be able to share knowledge while safeguarding competitive advantages. Finally, consumer engagement and readiness for sustainable change should be further analyzed, especially the mechanisms like crowdsourcing, living labs, and co-design platforms that integrate end-user perspectives.

8.8 Conclusion

According to research, open innovation (OI) provides an effective approach to overcoming challenges associated with the transition to a circular economy (CE) at all levels. By enabling collaboration and collective work across sectors, fostering OI ecosystems, and promoting knowledge exchange, OI supports the design and implementation of circular solutions at macro, meso, and micro levels.

At the macro level, OI mechanisms towards CE may include living labs, regulatory sandboxes, and open data practices. At the meso level, industrial symbiosis, innovation networks and partnerships, and blockchain and IoT for a sustainable supply chain are known OI practices that support and enable CE implementation. Finally, at the micro level, adapting a company's organizational structure, culture, and leadership plays a vital role in adopting open circular business models.

While the researched approach is promising, literature highlights several barriers and challenges, mainly at the meso and micro levels. Regarding the micro level, technical barriers – such as knowledge gaps and missing technologies – and cultural obstacles to openness are often mentioned. The openness paradox has also been recognized by research and practice as a barrier, which refers to intellectual property (IP) protection issues and the challenge of sharing while being protected from the competition. At the meso level, the misalignment of stakeholders is one of the most frequently observed challenges for networks and systems, while issues at the macro level include external socio-political and economic conditions, which often are accompanied by uncertainty.

Clear governance frameworks for IP protection and funding mechanisms are important to break the aforementioned barriers at both the macro and meso levels, as well as to establish circular supply chains. At the micro level, openness capacity-building practices, such as training workshops and innovation programs, will support organizations to adopt circularity via OI.

For policymakers, OI offers tools for participatory governance and adaptive regulation. It serves as a mechanism to unlock new business models and value chains for the industry. For academia, it presents fertile ground for cross-disciplinary exploration and applied impact. Realizing the full potential of OI in CE transition, however, requires strategic alignment, institutional support, and continued research into the enablers and constraints of open collaborative innovation.

References

Ahmadov, T., S. Durst, W. Gerstlberger, and E. Kraut, "SMEs on the way to a circular economy: insights from a multi-perspective review," *Manag. Rev. Q.*, vol. 75, no. 1, pp. 289–322, Feb. 2025. https://doi.org/10.1007/s11301-023-00380-2.

Almirall, E. and J. Wareham, "Living labs and open innovation: roles and applicability," *Electron. J. Virtual Organ. Netw.*, vol. 10, p. 95106, 2008.

Antikainen, M., A. Aminoff, O. Kettunen, H. Sundqvist-Andberg, and H. Paloheimo, "Circular economy business model innovation process – Case study," in *Sustainable Design and Manufacturing 2017, Smart Innovation, Systems and Technologies*, vol. 68, G. Campana, R. J. Howlett, R. Setchi, and B. Cimatti, Eds., Cham: Springer International Publishing, 2017, pp. 546–555. https://doi.org/10.1007/978-3-319-57078-5_52.

Bocken, N. and P. Ritala, "Six ways to build circular business models," *J. Bus. Strategy*, vol. 43, no. 3, pp. 184–192, Apr. 2022. https://doi.org/10.1108/JBS-11-2020-0258.

Bogers, M., H. Chesbrough, and R. Strand, "Sustainable open innovation to address a grand challenge: Lessons from Carlsberg and the green fiber bottle," *Br. Food J.*, vol. 122, no. 5, pp. 1505–1517, Jan. 2020. https://doi.org/10.1108/BFJ-07-2019-0534.

Brown, P., N. Bocken, and R. Balkenende, "How do companies collaborate for circular oriented innovation?," *Sustainability*, vol. 12, no. 4, p. 1648, Feb. 2020. https://doi.org/10.3390/su12041648.

Brown, P., C. Von Daniels, N. M. P. Bocken, and A. R. Balkenende, "A process model for collaboration in circular oriented innovation," *J. Clean. Prod.*, vol. 286, p. 125499, Mar. 2021. https://doi.org/10.1016/j.jclepro.2020.125499.

Chistov, V., N. Aramburu, and J. Carrillo-Hermosilla, "Open eco-innovation: A bibliometric review of emerging research," *J. Clean. Prod.*, vol. 311, p. 127627, Aug. 2021. https://doi.org/10.1016/j.jclepro.2021.127627.

Dautremont, C. and E. Gobbo, "Mapping of circular construction ecosystems' characteristics: interconnections, relationships, and synchronization of stakeholders at the micro, meso, and macro scales," *Sustainability*, vol. 17, no. 2, p. 541, Jan. 2025. https://doi.org/10.3390/su17020541.

De Jesus, A. and S. Mendonça, "Lost in transition? Drivers and barriers in the eco-innovation road to the circular economy," *Ecol. Econ.*, vol. 145, pp. 75–89, Mar. 2018. https://doi.org/10.1016/j.ecolecon.2017.08.001.

Dolgova, O. I. and A. Y. Nikitaeva, "Evaluation of the potential for the development of the circular industry in the region: A new approach," *Recycling*, vol. 10, no. 2, p. 38, Mar. 2025. https://doi.org/10.3390/recycling10020038.

Drofenik, J., T. Seljak, and Z. Novak Pintarič, "A multi-level approach to circular economy progress: Linking national targets with corporate implementation," *J. Clean. Prod.*, vol. 493, p. 144902, Feb. 2025. https://doi.org/10.1016/j.jclepro.2025.144902.

Egri, C. P. and S. Herman, "Leadership in the north American Environmental Sector: values, leadership styles, and contexts of environmental leaders and their organizations.," *Acad. Manage. J.*, vol. 43, no. 4, pp. 571–604, Aug. 2000. https://doi.org/10.2307/1556356.

Eisenreich, A., J. Füller, and M. Stuchtey, "Open circular innovation: How companies can develop circular innovations in collaboration with stakeholders," *Sustainability*, vol. 13, no. 23, p. 13456, Dec. 2021. https://doi.org/10.3390/su132313456.

Ellen MacArthur Foundation, "What is the meaning of a circular economy and What are the main principles?" *Introduction to Circular Economy*. [Online]. n.d. Available: https://www.ellenmacarthurfoundation.org/topics/circular-economy-introduction/overview?utm_source=chatgpt.com.

European Commission, *A European Strategy for Data*. European Commission, 2020. [Online]. Available: https://eur-lex.europa.eu/legal-content/EN/TXT/?uri=CELEX%3A52020DC0066

European Environment Agency, *With People and for People: Innovating for Sustainability.* LU: Publications Office, 2021. Accessed: Aug. 19, 2025. [Online]. Available: https://data.europa.eu/doi/10.2800/56160

Geerken, T., J. Schmidt, K. Boonen, M. Christis, and S. Merciai, "Assessment of the potential of a circular economy in open economies – case of Belgium," *J. Clean. Prod.*, vol. 227, pp. 683–699, Aug. 2019. https://doi.org/10.1016/j.jclepro.2019.04.120.

Geissdoerfer, M., P. Savaget, N. M. P. Bocken, and E. J. Hultink, "The circular economy – A new sustainability paradigm?," *J. Clean. Prod.*, vol. 143, pp. 757–768, Feb. 2017. https://doi.org/10.1016/j.jclepro.2016.12.048.

Harris, S., L. Staffas, T. Rydberg, and E. Eriksson, *Renewable Materials in the Circular Economy,* IVL Swedish Environmental Research Institute, 2018. https://doi.org/10.13140/RG.2.2.11296.43523.

Haskins, C., "Multidisciplinary investigation of eco-industrial parks," *Syst. Eng.*, vol. 9, no. 4, pp. 313–330, Dec. 2006. https://doi.org/10.1002/sys.20059.

Jakobsen, S., T. Lauvås, F. Quatraro, E. Rasmussen, and M. Steinmo, "Innovation for a circular economy," *SSRN Electron. J.*, 2021. https://doi.org/10.2139/ssrn.3857012.

Jesus, G. M. K. and D. Jugend, "How can open innovation contribute to circular economy adoption? Insights from a literature review," *Eur. J. Innov. Manag.*, vol. 26, no. 1, pp. 65–98, Jan. 2023. https://doi.org/10.1108/EJIM-01-2021-0022.

Kirchherr, J., D. Reike, and M. Hekkert, "Conceptualizing the circular economy: An analysis of 114 definitions," *Resour. Conserv. Recycl.*, vol. 127, pp. 221–232, Dec. 2017. https://doi.org/10.1016/j.resconrec.2017.09.005.

Köhler, J., S. D. Sönnichsen, and P. Beske-Jansen, "Towards a collaboration framework for circular economy: The role of dynamic capabilities and open innovation," *Bus. Strategy Environ.*, vol. 31, no. 6, pp. 2700–2713, Sept. 2022. https://doi.org/10.1002/bse.3000.

Lüdeke-Freund, F., S. Gold, and N. M. P. Bocken, "A review and typology of circular economy business model patterns," *J. Ind. Ecol.*, vol. 23, no. 1, pp. 36–61, Feb. 2019. https://doi.org/10.1111/jiec.12763.

Mankata, L. M., P. Antwi-Afari, and S. T. Ng, "A blockchain-based supply chain framework for advancing circular economy in the construction industry," *J. Clean. Prod.*, vol. 519, p. 145966, Aug. 2025. https://doi.org/10.1016/j.jclepro.2025.145966.

Martins, E. C. and F. Terblanche, "Building organisational culture that stimulates creativity and innovation," *Eur. J. Innov. Manag.*, vol. 6, no. 1, pp. 64–74, Mar. 2003. https://doi.org/10.1108/14601060310456337.

Nerone, Murilo Agio, Canciglieri Junior Osiris, and Liao Yongxin, "Classification of the open innovation practices: The creativity level," in *Advances in Transdisciplinary Engineering,* Amsterdam: IOS Press, 2014. https://doi.org/10.3233/978-1-61499-440-4-871.

Patala, S., A. Salmi, and N. Bocken, "Intermediation dilemmas in facilitated industrial symbiosis," *J. Clean. Prod.*, vol. 261, p. 121093, July 2020. https://doi.org/10.1016/j.jclepro.2020.121093.

Perotti, F. A., A. Bargoni, P. De Bernardi, and Z. Rozsa, "Fostering circular economy through open innovation: Insights from multiple case study," *Bus. Ethics Environ. Responsib.*, vol. 34, no. 2, pp. 390–408, Apr. 2025. https://doi.org/10.1111/beer.12657.

Rahmat, D. A., A. A. Rumanti, M. A. Pulungan, A. S. Rizaldi, and M. Amelia, "Evaluating the role of open innovation and circular economy in enhancing organizational performance: Insights from batik small and medium enterprises in Banyuwangi, Indonesia," *Sustainability*, vol. 16, no. 24, p. 11194, Dec. 2024. https://doi.org/10.3390/su162411194.

Skordoulis, M., S. Ntanos, G. L. Kyriakopoulos, G. Arabatzis, S. Galatsidas, and M. Chalikias, "Environmental innovation, open innovation dynamics and competitive advantage of medium and large-sized firms," *J. Open Innov. Technol. Mark. Complex.*, vol. 6, no. 4, p. 195, Dec. 2020. https://doi.org/10.3390/joitmc6040195.

Suchek, N., C. I. Fernandes, S. Kraus, M. Filser, and H. Sjögrén, "Innovation and the circular economy: A systematic literature review," *Bus. Strategy Environ.*, vol. 30, no. 8, pp. 3686–3702, Dec. 2021. https://doi.org/10.1002/bse.2834.

Surya, D., R. K. Rahim, M. Hamsal, S. Candra, and W. Gunadi, "Mapping the nexus between open innovation and circular economy: a decade of bibliometric evidence," *Front. Sustain.*, vol. 6, p. 1638284, July 2025. https://doi.org/10.3389/frsus.2025.1638284.

Triguero, Á., M. C. Cuerva, and F. J. Sáez-Martínez, "Closing the loop through eco-innovation by European firms: Circular economy for sustainable development," *Bus. Strategy Environ.*, vol. 31, no. 5, pp. 2337–2350, July 2022. https://doi.org/10.1002/bse.3024.

Yoshino, M., B. Sadlek, M. Yarime, and A. Ali, "Knowledge absorption pathways for eco-innovation: an empirical analysis of small and medium-sized enterprises in the European Union," *Eur. J. Innov. Manag.*, vol. 28, no. 2, pp. 426–453, Jan. 2025. https://doi.org/10.1108/EJIM-02-2023-0136.

Zhang, D., X. Huang, Y. Wen, P. Trivedi, and S. Joghee, "Sustainable circular business model for transparency and uncertainty reduction in supply chain management," *J. Theor. Appl. Electron. Commer. Res.*, vol. 16, no. 4, pp. 959–975, Mar. 2021. https://doi.org/10.3390/jtaer16040054.

9

Should Innovations in the Fields of Energy and Environment Envisage a Low-Carbon Economy?

Konstantinos G. Aravossis, Grigorios L. Kyriakopoulos, Ioannis Sebos, and Yassine Charabi

9.1 Introduction

In our better understanding of the contentious issue of the transformation of the urban distribution industry, the self-distribution mode was introduced in the form of low-carbon urban distribution mode. This mode determined the main research types of goods, being a methodological proposal to the specific enterprises. Focusing on climate change and the target of energy conservation and emission reduction, the proposed method of urban distribution mode can reduce carbon emission intensity, identify core influencing factors of urban distribution modes, and promote noticeable reduction of carbon emissions intensity, showing that there are methods described in the literature that can effectively guide enterprises to develop low-carbon urban distribution mode and reduce carbon emission intensity of urban distribution. Similar to urban distribution is the issue of smart grid construction, being an important carrier technology to effectively promote the development of low-carbon economy. Considering that the demand response (DR) is commonly regarded as an important core technology in the smart grid field, flexible and interactive features of the core business in smart electricity can be demonstrated. Aiming to achieve the interoperability between the grid side and the user side, there has been business conducted in smart grid user interface standards that further support demand-side management. The ultimate goals are the highest benefits of energy conservation and emissions reduction resulting from renewable power generation through optimizing the annual operation of the power system. Subsequently, the model's validation and verification of the simulation results in practical regional power systems and corresponding solving algorithms.

DOI: 10.1201/9781003654889-9

From an environmental perspective, and more noticeably in recent decades, greenhouse gases (GHGs) are responsible for global damage, global warming, and climate change. The concurrent problems in various sectors of the world economy make an imperative need the reduction of the negative impact on the environment and the optimum management of GHGs emissions in the regional economy. Therefore, the best management methodologies should be envisaged towards reducing total costs for economic entities in the interests of the entire region. The implementation of the methodologies are tools of stimulating economic entities, the use of which will realistically bring them and the nearby society closer to a low-carbon economy.

On the other hand, there are gaps at local, national, and regional levels concerning policymaking and project formulation related to climate change, sustainable development, and human security. It is noteworthy that whenever an increased level of regional growth is prospected, this region then needs to tread down a path towards a low-carbon economy in order to face up to the challenges of climate change. Contrarily, inadequate regional cooperation, energy insecurity, and weak governance are identified as key constraints to achievement of a low-carbon economy at the national level of analysis. The most noticeable human actions and priorities related to confronting climate change are the adoption of specific climate change policies, the meeting of forums for regional cooperation, and private sector initiatives which are able to promote sustainable practices. The transformation and scaling up from policy planning to action needs to be undertaken, while the main barriers of such a transformation are political instability, a lack of economic integration, and limited natural resources that need to be acknowledged and surmounted prior to achievement of sustainable development in regional scale of analysis. A plethora of literature review studies disclosed numerous examples and mechanisms that may encourage coordinated and integrated policies and mechanisms towards low-carbon economy, underscoring the effective response to climate change and the challenges facing regional/topical interest.

Among numerous examples and mechanisms of encouraging coordinated and integrated policies noted in the literature, the authors of this study developed a research method to collect and include timely published studies on jointly recognizing sectors emerging in the literature and low-carbon economy in local, regional, national, or continental interest. The applied methodology therefore contains a collection of studies collected from a literature search at the middle of the year 2025, covering documents with "low[-] carbon economy" in their titles. This literature search described in Section 9.2 enabled the organization of search results under the following literature-defined areas: "low[-]carbon economy" coupled with the search terms "energy" and "environment," accordingly. The steps of literature collection and organization in clusters has been conveyed in Section 9.3, while the main priority actions and utmost significant issues per cluster are demonstrated in the Section 9.4. The chapter concludes with Section 9.5.

9.2 Literature Review

9.2.1 Theoretical Overview of the Joint Fields of "Low-Carbon Economy" and "Energy"

The implementation of the principle of low-carbon development is in line with the gradual transformation of the world's energy – the transition from fossil fuels as the main source of primary energy resources to other sources of energy. The relevance of the study is determined by the fact that the most important element of development is the possibility of consumption of energy resources and, accordingly, the possibility of their production in the context of economic changes (Kapitonov, 2021). In the relevant literature, researchers have defined that there was an ongoing interest of energy sources of production in global positive dynamics during the period from 2008–2017. This period has been characterized as a transition era in the field of fossil fuels elimination and the gradual promotion of alternative energy. This transitional phase stimulated its development through state regulation of the industry, subsequently improving the investment and innovation climate of nations as a whole. While the issue of the sustainable development structure of energy security has been raised for the first time during the last decade, it can be concluded that in the near future, trends will be maintained not only in the scales of methods, statistical analyses, and mathematical models under multi-criteria methods of estimates, but also in real-world and large-scale energy production applications/units/plants (Kapitonov, 2021).

Among the most widely known mathematical models is that of game theory for information analysis (Wang, 2022). In such a study, evaluation of the environmental accounting information disclosure level of the manufacturing industry in the low-carbon economy environment was applied through adopting game theory for information analysis in resolving the problem of environmental accounting of the Chinese manufacturing industry. Building up a hybrid strategy static game model for government regulators and manufacturing enterprises can be developed, and the effectiveness of the proposed model can be verified by experimental results (Wang, 2022).

9.2.2 Theoretical Overview of the Joint Fields of "Low-Carbon Economy" and "Environment"

From an economics perspective, it is scientifically challenging to analyze low-carbon economy in alignment with the optimization and distribution of a new financial resource named carbon emission rights to produce profound effect on enterprise operation and decision-making (Xie et al., 2012). Such a type of analysis includes cost structure, consumption concept, profit model, and market risk. The critical point here is the integrating approach and promotion of rational proposals including seeking the law of demand, making reasonable price decisions, strengthening cooperation between enterprises, and focusing on the carbon price, which has important impact on the enterprise's cost and profit (Xie et al., 2012).

From a social perspective, the rapid growth of social and economic development has brought great challenges to the development of regional resources as it relates to the environment. The coordinated development of resources, environment, and economic growth is not only the engine of industrial growth but also an accelerator to promote industrial transfer and optimal allocation of resources, which has injected a strong impetus into social development (Wang, 2023). However, with the continuous upgrading of the industrial infrastructure, environmental pollution, and unbalanced economic development, resource shortages and other problems have gradually emerged in various important development areas (Wang, 2023).

With the deepening of the concept of sustainable development, the coordinated development of regional resources, environment, and economic growth has attracted more and more attention. Among the main contemporary contributors of sustainable development is the low-carbon economy, which has strong feasibility for promoting the coordinated development of regional resources, environment, and economic growth (Wang, 2023). Regarding the sustainable development of regional resources, environment, and economic growth, it is noteworthy that in recent times, the ongoing development of industrialization, the concurring ecological and environmental problems, and the environmentally burdening practices of direct burning of crop straw and livestock manure also cause environmental depletion of the healthy ecological environment. In order to promote sustainable development of the ecological environment, low-carbon economy was demonstrated as a tool to integrate the environmental management of enterprises through establishing the decision-making model of ecological environment management and carrying out related research (Li and Gao, 2023). In this study, it was shown that the system dynamics method can effectively reduce the section water quality prediction and pollutant discharge prediction errors within the management decision-making model. Improving the accuracy of the model prediction of managing agricultural waste of high contents of convertible biomass energy can provide a large amount of renewable energy for humanity, especially in local and remote areas, and in mainland electricity grids (Li and Gao, 2023).

Closing the literature review, it is also noteworthy that in the last five years of publication, researchers attributed an economic development model of low-carbon economy based on energy-efficient utilization and clean development, and characterized this economy by low energy consumption, low pollution, and low emissions (Lingjun and Chao, 2021). Low carbon can be collectively conceptualized under different terms at different situations and research contexts, like that of: (a) low-carbon hospital which is a new concept derived from low-carbon economy and refers to the operation process of the implementation of low energy consumption, low material consumption, low financial consumption, low emissions, land ow pollution (Lingjun and Chao, 2021);

or (b) interior design, which is a key strategy for boosting home comfort, thus greatly affecting both the efficiency of interior decorative design and the level of energy usage, focusing on resolving a variety of household problems including illogical design principles, poor material selections, high energy consumption of design schemes, and a failure to adhere to modern low-carbon design principles. In this context, the idea of low-carbon design is another course of private interior design, underlining the environmental security, usefulness, and logical interior design (Li and Haron, 2024).

9.2.3 Thematic Contexts of Low-Carbon Economy in the Contexts of Energy and Environment

The collected documents of the conducted literature search enabled us to organize all yielded documents into the following four contexts, each having all relevant publications presented in the form of the following four tables.

Table 9.1	Artificial Intelligence (AI) Context
Table 9.2	Geographical-Spatial Context
Table 9.3	Projects Context
Table 9.4	Models and Analytical Techniques Context

The number of citations per context-category and the relevant weighted significance in the total of the published studies that count for the co-terms "low-carbon economy," "energy," and "environment" are presented in Table 9.5.

Based on Tables 9.1–9.4 and the collective data of Table 9.6, it can be concluded that there is a steady increase of literature production in the field

TABLE 9.1

Artificial Intelligence (AI) Context

Artificial Intelligence (AI) Context (in Alphabetical Order of AI Areas Named)	Key Aspects	Reference Item
Energy strategies	Viewing the transition to a low-carbon economy in the light of advancing global economic policies and hydrogen production through energy strategies that are based on harnessing artificial intelligence	Hou et al. (2024)
Smart energy industry	The pivotal roles of artificial intelligence and low-carbon economy to support China's smart energy industry and the broader national sustainable development	Shi et al. (2022)

TABLE 9.2

Geographical-Spatial Context

Country/Region/ Spatial Context (Alphabetical Order of Countries Named)	Key Aspects	Reference Item
Australia	Evaluating Australia's energy regulation landscape regarding: (a) characterization of strengths and weaknesses in Australia's current energy regime; and (b) recommendation of encouraging policies to facilitate Australia's transition towards low-carbon economy	Gunningham and Bowman (2016)
China	These studies explore the shift in the Chinese government's policies towards a low-carbon economy by adopting more environmentally friendly solar energy. In China, coal is a dominant component of energy mix. This energy mix is expected to remain the same over the next 30–40 years. The energy challenges China faces include the mitigation processes to be gradually transformed into the conceptual scheme: "IGCC + polygeneration + CCUS." The synergistic utilization of coal with other energy resources can be prosperous, counting also the scientific and the technological problems to achieve these synergies. These studies denote that China is an example to investigate the green finance policy, jointly with the effect of fossil energy use rights trading (FET), renewable energy certificates trading (RECT), and renewable energy subsidies (RES) towards the low-carbon economy using a dynamic recursive computable general equilibrium (CGE) model. The utilization of the super-efficiency slack-based measure (SBM) model in order to measure the energy efficiency of 30 provinces in China, conducting the empirical orthogonal function (EOF) aiming at analyzing its spatial-temporal evolution.	Gulzar et al. (2020), Ni and Chen (2011), Jiang et al. (2010), Pan and Dong (2023), Zhang et al. (2020), Chen and Li (2024), Shen and Sun (2016), Wang and Huang (2022), Li et al. (2021), Hou et al. (2011), Zhang and Wang (2025), Sun and Miao (2022)
Bangladesh	The study focused on the role of clean energy transition for achieving a low-carbon economy	Murshed et al. (2021)
Bulgaria	The study focused on the transition to a low-carbon economy in the energy sector and how this transition effects on the economic growth of Bulgaria	Zlatinov et al. (2019)
Cuba	The study focused on the low-carbon economy and biomass-based electricity	Sagastume Gutiérrez et al. (2018)

(Continued)

TABLE 9.2 (*Continued*)

Country/Region/ Spatial Context (Alphabetical Order of Countries Named)	Key Aspects	Reference Item
Germany	The target of low-carbon economies has been examined in alignment with the purposes of reindustrialization through development of stakeholder-based scenarios for energy-intensive industries, having also the strategic plan of national energy production following the framework of green economy, low-carbon economy, socio-environmental conflicts and equitable society	Lechtenböhmer et al. (2015), Weber and Cabras (2017)
Japan	Low-carbon economy has been perceived and investigated through energy saving and business model, as well as through nuclear energy transition	Watanabe and Nishimura (2013), Ishida (2018)
Kazakhstan	An approximate relationship analysis, a decoupling relationship analysis, and a trend analysis to explore the relationship between energy consumption and economic growth of Kazakhstan using data for the period of 1993–2010; the proposition of strategies included the improvement of energy consumption structure, development of new energy and renewable energy, utilization of cleaner production technologies, adjustment and optimization of the industrial structure, and expanding national forest areas	Xiong et al. (2015)
Pakistan	The symmetric and asymmetric impacts of green energy, eco-innovation, and the urbanization in explaining a low-carbon economy for Pakistan	Ullah et al. (2023)
Poland	The study of Slupik et al. (2021) explored the energy-related behaviour of consumers from the Silesia province (Poland) towards a low-carbon economy. The study of Dembicka-Niemiec et al. (2023) was primarily concentrated with identifying the scope of the use of European Union (EU) funds for the formation of a low-carbon economy by enterprises providing energy services in Poland during 2014–2020. These authors pointed out that EU funds were important in the financing of investments by the largest energy companies to decarbonize the economy. The research analysis also showed that the surveyed companies were pursuing the goals of Directive 2012/27/EU of the European Parliament and of the Council for Energy Efficiency, primarily concerning a low-carbon economy.	Słupik et al. (2021), Dembicka-Niemiec et al. (2023)

(Continued)

TABLE 9.2 (*Continued*)

Country/Region/ Spatial Context (Alphabetical Order of Countries Named)	**Key Aspects**	**Reference Item**
South Africa	These studies focused on shifting from an energy-intensive to low-carbon economy vision. This vision put in place in order to mitigate negative impacts of climate change, address the energy mix, and move towards a low-carbon economy. The policy landscape at various levels namely is indicatively composed by acts of parliament, policies (white papers and green papers), strategies, and action plans, as well as regulations and instruments. It was shown that South Africa has moved swiftly to establish an effective legislative regime for renewable energy investment, which is assisting the country in fulfilling its vision, as outlined earlier.	Winkler and Marquand (2009), Nhamo and Ho (2011)
Spain	Key differences between Spanish regions are identified in the way the renewable energy implementation process has been carried out. The research brings these different characteristics into focus and analyses the strengths and weaknesses of the renewable energy implementation process in each region. Very powerful actors (e.g., multinational energy companies or governments) can dominate the process, and thus, systemic instability can be reduced.	Alonso et al. (2016)
Thailand	The study compares approaches to energy transition in the light of policies of enhancing Thailand's vision of a low-carbon economy.	Kanchana (2024)
United Kingdom	A vision of low-carbon economy through the dimensions of space and place of energy transition achievement, the study focuses on the UK government's policy for a low-carbon transition.	Bridge et al. (2013)
United States of America	For 40 years, tax subsidies have "paid" private financial players to invest in renewable energy by allowing them to claim legal tax shelters against their other income. In this political economic analysis, the raised questions are: first, how U.S. renewable energy acquired this peculiar form of public finance through the tax code, unique in the global industry; and second, exploring how the model has shaped U.S. renewables financing, development, and ownership. Two decisive moments were the California "wind rush" in the 1980s, and the ongoing renewables boom of the last 15 years.	Knuth (2023)

(Continued)

TABLE 9.2 (*Continued*)

Country/Region/ Spatial Context (Alphabetical Order of Countries Named)	Key Aspects	Reference Item
	Via tax equity, a handful of major U.S. banks dominate financing for renewables and other politically embattled public goods. These banks exert a troubling ability to extract rents for their capital, gatekeep what projects get built and by whom, and stall U.S. renewables development altogether. Nowadays, the practice is increasingly strained by similar problems, including growing public costs, private capacity ceilings, and amplification of sectoral crises.	
Clusters/groups of two or more countries co-evaluated	A) United Kingdom and the United States of America: Global warming phenomenon through replacing natural gas with hydrogen in the domestic energy sectors of future low-carbon economies B) Spain and the EU28: The advancement of an energy-efficient low-carbon economy, through an input–output logarithmic mean divisa index (LMDI) decomposition of carbon dioxide emissions C) G7 and BRICS: Improving low-carbon economies of G7 and BRICS countries through analysis of energy efficiency and environmental Kuznets curves D) OECD countries: Viewing the transition of low-carbon economy through implementing the wider diffusion of renewable energy, globalization, and higher economic growth E) EU and China: The energy policy of the EU and China towards the Arctic, in alignment with falling oil and gas prices, climate change, and low-carbon economies F) European context: Investigating to what extent European renewable energy mutual funds could foster the transition to a low-carbon economy	A) Field and Derwent (2021) B) Serrano-Puente (2021) C) Chang (2015) D) Ali et al. (2024) E) Giber (2019) F) Martí-Ballester (2019)

TABLE 9.3

Projects Context

Project Context	**Key Aspects**	**Reference Item (from Oldest to Newest, and in Alphabetical order of First Author's Last Name)**
The binary coded quantum-behaved particle swarm optimization (QPSO) algorithm is used to perform simulation and analysis in an IEEE six-node system, and a power grid planning scheme	Establishing a reasonable power grid is essential for promoting the development of new energy. Due to the intermittent and non-dispatchable nature of solar and wind-generated electricity sources, planning conventional power grids becomes more complicated. The proposed model supported the development of a low-carbon power grid.	Yang and Niu (2018)
Three scenarios: business-as-usual (BAU), high industry growth (HIG) and renewable energy export (REE)	Energy policy scenarios to transition to a low-carbon economy by 2050 in the case of the Northern Territory (NT) of Australia. The effectiveness of the energy transition policies is determined by careful consideration of the unique circumstances and existing barriers in the NT. While the future could be a combination of HIG and REE scenarios, there should be a groundwork for a smoother transition to low-carbon economy at the NT level.	Hosein et al. (2023)
Optimization of financial performance among energy enterprises	A set of financial evaluation index systems of energy industry enterprises was evaluated under the background of a low-carbon economy from four dimensions: profitability, asset quality, debt risk, and business growth. The conduct of analytic hierarchy process (AHP) was able to measure the comprehensive contribution of financial indicators of low-carbon production and operation. The proposed system was based on a financial management decision-making index to quantify low-carbon operation of energy enterprises. This is an important guide of energy enterprises to reduce costs and to increase the efficiency and low-carbon operation in practice.	Li et al (2024)

(Continued)

TABLE 9.3 (*Continued*)

Project Context	Key Aspects	Reference Item (from Oldest to Newest, and in Alphabetical order of First Author's Last Name)
Bioenergy, geothermal, solar, hydropower or hydrogen, ocean, and wind (BIGSHOW) project	The perspectives and the prospects of low-carbon economy through renewable energy and net-zero goals. Sound evidence was shown for speeding up the shifting progress of a precarious fossil fuel-based economy towards a sustainable low-carbon one, in which BIGSHOW has played a known protagonist role.	Nguyen et al. (2024)
Virtual power plants (VPPs); dynamic cost optimization; mixed-integer linear programming (MILP) model; CPLEX solver for cost-effective decision-making	A polymorphic energy coordination strategy was developed following a range of case studies, which demonstrate the strategy's effectiveness that balance both economic and environmental objectives through multiple operational scenarios; in this study, there is also an experimental validation that confirmed its practical advantages over conventional approaches. Critical insights for policymakers and energy enterprises should guide a stable framework for low-carbon transition.	Zheng et al. (2025)
Miscellaneous and unclassified projects	There is a broad thematics of coverage of low-carbon energy projects, covering an international context of topics, including: marine energy, transformational materials, energy-efficient integrated energy system considering carbon capture-two stage power-to-gas (P2G) hydrogen coupling, bio-renewables in regional energy and transportation systems, and dispatching of integrated energy system with P2G-HGT coupling wind power absorption, as well as the role of renewable energy and natural gas, all towards low-carbon energy applications/perspectives	Kitasei (2010), Robinson (2010), Krishnan and McCalley (2016), Chen et al. (2019), Kuznetsov and Gasenko (2023), Liu et al. (2023), Zhu et al. (2023), He et al. (2024), Zhang and Kuang (2024)

TABLE 9.4

Models and Analytical Techniques Context

Context of Models and Analytical Techniques	Key Aspects	Reference Item (from Oldest to Newest, and in Alphabetical order of First Author's Last Name)
Chemical polishing wastewater under the principles of low-carbon economy	By applying a combination of advanced analytical techniques of SEM-EDS, XRD, FITR, and Mössbauer analyses, authors targeted at pollutants removal, resource recovery, and energy generation. The calculation of the total output energy reached 0.33 kW h/t wastewater, with maximum output voltage at 0.19 V. The study provided a novel method of potential economic advantages to support sustainable development of chemical polishing wastewater treatment.	Ding et al. (2023)
Cement manufacturing industry, under the principles of circular economy to produce low-energy cement	Analysis was performed via a variety of advanced techniques, including X-ray fluorescence (XRF), X-ray diffraction (XRD), attenuated total reflectance Fourier trans-form infrared spectroscopy (ATR-FTIR), thermal analysis (TG-DTG-DSC) scanning electron microscopy, and energy dispersive X-ray spectroscopy analysis (SEM-EDS). The samples have been prepared in order the effect of the addition of different industrial waste types, as a substitution for raw materials, in the production of low-energy cement with high dicalcium silicate content to be examined. Waste was utilized as a raw material for the production of a more useful substance, considering and adopting the fundamental principles of the circular economy.	Martínez-Martínez et al. (2023)
Life cycle analysis (LCA) towards dual transformation of economy and energy following the principles of low-carbon economy	An accurate accounting of carbon emissions was the first step of this analysis. Then, Internet of Things (IoT) technologies were employed to collect real-time data with confidence to monitor the environmental consequences of products throughout the life cycle analysis (LCA). Energy efficiency, energy structure optimization, and market-based economic measures such as energy/carbon taxes have minimized energy consumption and carbon emissions under a long-term low-carbon strategy.	Huang et al. (2024)

(Continued)

TABLE 9.4 (*Continued*)

Context of Models and Analytical Techniques	Key Aspects	Reference Item (from Oldest to Newest, and in Alphabetical order of First Author's Last Name)
Regional energy consumption policy practice of regional interest (in the Shaanxi Province of China)	There is an influential role of policy variables targeting at the evolution of energy consumption structure by constructing a system dynamics model. In the study, it was shown great significance to the optimization of energy structure led by coal. The constructed and applied model can effectively simulate and adjust practical policies, guiding a reference tool for the development of low-carbon economic policies of broader regional utility.	Sun and Onuh (2024)
Multi-algorithm optimization in the case study of a province in China aiming to accelerate the high-quality development of green and low-carbon economy between a regional low-carbon economy and the energy structure	The development of a comprehensive evaluation index system was targeted to link between low-carbon economy and energy structure based on the spatial data infrastructure (SDI) conceptual framework. Using the entropy weight method for weighting and calculating the development index in the period 2014–2021 was employed to measure the level of regional development. Subsequently, using the coupling coordination model can measure the level of regional coupling. It was proved that the overall development level of low-carbon economy showed an upward trend. Moreover, the coupling development of low-carbon economy and energy structure in the Chinese province was on an ongoing rise. It was also noteworthy that the overall resistance in the dimension of energy structure implied a constant growth trend, while the overall resistance in the dimension of low-carbon economy was an indicator of a steady downward trend.	Zhao et al. (2024)

(*Continued*)

TABLE 9.4 **(*Continued*)**

Context of Models and Analytical Techniques	Key Aspects	Reference Item (from Oldest to Newest, and in Alphabetical order of First Author's Last Name)
Investing in low-carbon economy and the role of energy efficiency in China	Recently, rapid economic growth in China accelerated the energy consumption. However, there is still low energy efficiency, making highly important research on improving the energy efficiency. In response to this problem, low-carbon economy necessitates the transformation from a currently high energy consumption through optimizing energy structure and increasing the energy efficiency. This study utilized an endogenous growth model to identify the effect on low-carbon economy growth. Then, a vector auto-regression model tested how financial development can effect on energy efficiency. It was argued that financial development can play a pivotal role in improving energy efficiency in the long run.	Zhou (2014)
Energy productivity and low-carbon economies	The joint applicability of the modified ordinary least squares (FMOLS) and canonical cointegrating regression (CCR) estimation methods and dynamic ordinary least squares (DOLS) for robustness disclosed that energy productivity discourages the use of renewable energy among European countries, having a negative effect on the utilization of said energy estimated at a decrease of renewable energy use by 0.24% for each 10% increase in energy productivity. The positive correlation between energy productivity and environmental sustainability arises from a reduction in energy inputs within the broader productivity framework. The transition towards greener energy landscapes initiates a strategic process of substituting conventional fuel sources with renewable alternatives, fostering environmental enhancement through energy source diversification.	Soto et al. (2025)

(Continued)

TABLE 9.4 (*Continued*)

Context of Models and Analytical Techniques	Key Aspects	Reference Item (from Oldest to Newest, and in Alphabetical order of First Author's Last Name)
Miscellaneous and unclassified models and analytical techniques	A corpus of non-classified models and analytical techniques have been focused on investments in renewable energy sources and a comparison between carbon economies through green technologies, the roles of energy efficiency and low-carbon economy investments to support sustainability development, the effects of energy saving and emission reduction of fossil energy based on low-carbon economy and the accompanying consumption structure optimization, a scenario prediction model for carbon emission calculation method, the applicability of neural network model for energy low-carbon economy and financial risk, the role of low-carbon economy for sustainable development of energy security, statistical analysis coupling with mathematical modelling, and the method of criteria estimates	Romano and Scandurra (2011), Zhou (2014), Li and Dai (2019), Yu and Huafeng (2019), Kapitonov (2020), Porshnov (2022), Alsiyabi et al. (2024), Li and Khan (2024), Liu et al. (2024), Gao (2025), Tariq et al. (2025), Wu (2025)

TABLE 9.5

Overview of the Yielded Documents per Context, Decade, Weighted Percentage

Context (in Descending Order)	Number of References (in Parentheses the Decades of Publication: 2020s, 2010s, 2000s)	Percentage (%)	Chronological Order in Decades (in Parentheses, %)
Geographical/spatial	35 (17, 17, 1)	50.00	2000s: 1 (1.43%) 2010s: 27 (38.57%) 2020s: 42 (60.00%)
Models and analytical techniques	19 (15, 4)	27.14	
Projects	14 (8, 6, 0)	20.00	
Artificial intelligence	2 (2, 0, 0)	2.86	
Total of 4 contexts	**70**	**100**	**70 (100)**

Source: Authors' synthesis

of low-carbon economy – having, however, a declining publication trend in the field of projects and an increasing trend in the field of AI from the first decade of the 2000s onwards. Half of total publication is devoted to the context of geographical/spatial areas, followed by the context of models and analytical techniques and a lower participation of studies in projects and AI. We cannot overstate that the decade 2010s has been already ended, while the current decade of the 2020s has four years yet to be completed. Therefore, it is anticipated that the contexts of geographical/spatial areas, as well as the contexts of models and analytical techniques, will continue to be characterized with an increase pace of publications in the short-run, followed by the fast emerging AI context which could surpass (or be synergistic with) that of the projects context.

9.3 Methodology and Results

The two basic criteria of extracting the studies referred to "low-carbon economy" were the search-terms of "energy" and "environment." In this context, the domains in which the relevant yielded documents have been allocated, where that of Scopus-filed results in the following six domains.

- Subject areas
- Source titles
- Clusters of key words
- Affiliations
- Countries/territories
- Funding sponsors

In this context, the "low-carbon economy" and "energy" results are presented in Tables 9.6–9.11, while "low-carbon economy" and "environment" results are presented in Tables 9.12–9.16, accordingly. The synthesis of yielded documents are totalled at the end of each table, summing to 100% and having the corresponding percentage-fractions presented in parentheses within the 12 tables. An indexing provision was also taken such that conceptually familiar and overlapping terms of: "key words" and "subject areas" were not to be kept detached or in isolation from each other, but rather to be organized together into broader and fully developed clusters per domain. This is the reason for gathering more than one term of conceptual affinity in some entries of these tables. The time of publishing the yielded documents is dating back to the last three decades of publication, covering the period of the last 25 years, from 2000 onwards.

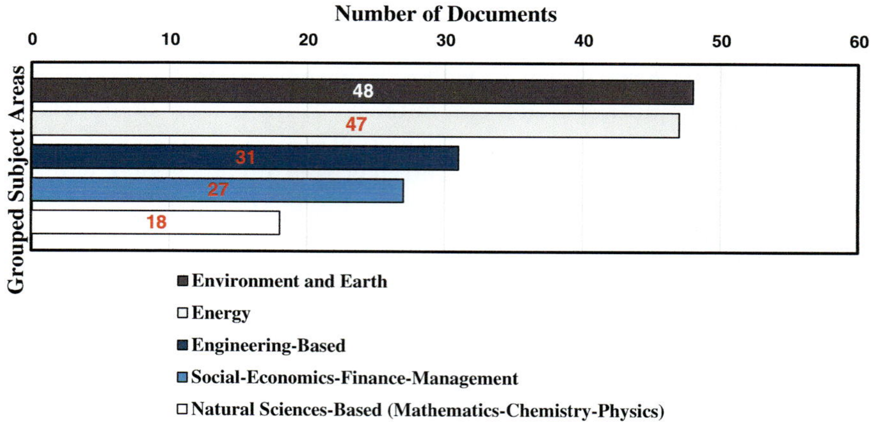

FIGURE 9.1
Top five grouped "subject areas" in the joint fields of "low-carbon economy" and energy.

Source: Authors' own synthesis

TABLE 9.6

Composition of the Top Five Grouped "Subject Areas" in the Joint Fields of "Low-Carbon Economy" and Energy

Subject Areas, #	Number of Documents in Descending Order, Number of Documents (%)
Environment and Earth, 48	Environmental science, 40; Earth and planetary sciences, 6; agricultural and biological sciences, 2 (28.1%)
Energy, 47	Energy, 47 (27.5%)
Engineering-based, 31	Engineering, 29; chemical engineering, 2 (18.1%)
Social-economics-finance-management, 27	Social sciences, 12; economics, econometrics, and finance, 10; business, management and accounting, 5 (15.8%)
Natural sciences-based (mathematics-chemistry-physics), 18	Mathematics, 13; chemistry, 3; physics and astronomy, 2 (10.5%)
Total, 171	**100**

Source: Authors' own synthesis

TABLE 9.7

The Top Six Source Titles in the Joint Fields of "Low-Carbon Economy" and Energy

Source Titles, #	Number of Documents in Descending Order (%)
Energies, 8	30.8
Journal of Cleaner Production, 5	19.3
Renewable Energy, 4	15.4
International Journal of Energy Economics and Policy, 3	11.5
Environmental Science and Pollution Research, 3	11.5
Energy, 3	11.5
Total, 26	**100**

Source: Authors' own study

TABLE 9.8

The Top Five Clusters of Key Words in the Joint Fields of "Low-Carbon Economy" and Energy

Clusters of Key Words	Number of Documents in Descending Order (%)
Carbon, 150	Carbon, 31; carbon emission, 15; emission control, 14; carbon dioxide, 11; carbon emissions, 10; global warming, 7; greenhouse gases, 6; greenhouse gas, 5; low-carbon, 5; carbon trading, 5; low-carbon, 4; CO_2 emissions, 4; carbon sequestration, 4; carbon capture, 4; low emission, 3; gas emissions, 3; coal, 3; carbon capture and utilization, 3; zero-carbon, 2; low-carbon technologies, 2; low-carbon economic, 2; low-carbon development, 2; carbon reduction, 2; carbon footprint, 2; a-carbon, 1 (38.8)
Low-carbon economy, 76	Low-carbon economy, 46; low-carbon economy, 30 (19.6)
Renewables, 66	Renewable energy resources, 23; renewable energy, 18; renewable energies, 8; renewable resource, 5; natural gas, 4; renewable energy consumption, 3; hydrogen, 2; hydrogen production, 2; alternative energy resources, 1 (17.0)
Economics, 54	Economic development, 15; economic growth, 11; economics, 8; environmental economics, 5; economic growth, 4; operating costs, 3; fuel economy, 2; financial development, 2; decision-making, 2; cost–benefit analysis, 2 (14.0)
Environment, 41	Environmental management, 15; sustainability, 9; sustainable development, 8; environmental sustainability, 3; environmental technology, 2; environmental pollution, 2; environmental factor, 2 (10.6)
Total, 387	**100**

Source: Authors' own study

TABLE 9.9
The Top 19 Affiliations in the Joint Fields of "Low-Carbon Economy" and Energy

Affiliations, #	Number of Documents in Descending Order (%)
Chinese Academy of Sciences, 7	21.5
University of Chinese Academy of Sciences, 5	15.4
Tsinghua University-Institute of Geographic Sciences and Natural Resources Research, 3	9.1
Zhejiang University, 2	6.0
The Australian National University, 2	6.0
Technical University of Denmark, 2	6.0
Taiyuan University of Technology, 2	6.0
Beijing Institute of Technology, 2	6.0
Nankai University, 2	6.0
University of Birmingham, 2	6.0
North China Electric Power University, 2	6.0
Plekhanov Russian University of Economics, 2	6.0
Total, 33	**100**

Source: Authors' own study

Further investigation of the Chinese sectors on which the research interest has focused is referred to in the study of Liang and Xu (2024), who established an evaluation framework for institutional coordinated growth across eight regions of China. These authors examined the low-carbon economy, agricultural products, and the resource environment as three interconnected subsystems, reporting that the agricultural product system captures 99.502% of its information, while the resource-environment system accounts for 84.823%, demonstrating robust explanatory power (Liang and Xu, 2024). From a normative and regulatory historical context in China, the national economic–agricultural–resource–environment (EARE) system progressed from sub-coordinated growth (2010–2014) to coordinated growth (2019–2020), moving from mild imbalance to high-quality growth. While at first the resources growth lagged behind economic development (2010–2015), a shift occurred to economic growth lagging behind resources and environmental growth (2015–2020). The critical point of this behaviour is the need for targeted policies to enhance regional sustainability and balanced development (Liang and Xu, 2024).

Based on Table 9.10, it was shown that the pace of renewable energy penetration in the international market is not synchronized or accelerated to all countries. For example, both China and Russia are among the most energy- and carbon-intensive countries in the world. The high level of technical abrasion and a low level of investments into modernization of their national energy industries cause huge energy waste and carbon emissions. Subsequently, this

TABLE 9.10

The Top Seven Countries/Territories in the Joint Fields of "Low-Carbon Economy" and Energy

Countries/Territories	Number of Documents in Descending Order (%)
China, 44	57.9
United Kingdom, 8	10.5
Spain, 7	9.2
Russian Federation, 5	6.5
United States, 4	5.3
India, 4	5.3
Germany, 4	5.3
Total, 76	**100**

Source: Authors' own study

is further regarded by countries relying on energy imports from these two mega-economies, creating an increasing threat to the security of supply and a major barrier to global climate change policy (Garbuzova and Madlener, 2012).

Actually, the problem is not only environmental or purely economic, but the focus is laid on investigating the progress and the future potential of the market-oriented mechanisms of joint implementation (JI) and green investment scheme (GIS) being considered as two possible channels for foreign direct investment (FDI) in transnational energy efficiency and carbon mitigation projects (Garbuzova and Madlener, 2012). In the context of fossil-based economies, the research focus can be directed to the current and to the future national energy efficiency and the GHG mitigation policies. In such a study, it was concluded that both JI and GIS are confronted with numerous barriers in their national energy markets. These authors also scrutinized the ability of energy service companies (ESCOs) as one of the market intermediary models to overcome some of emerging barriers in the process of effectively integrating JI and GIS into their long-term business strategies (Garbuzova and Madlener, 2012). Due to the compatibility of the main features of JI and GIS with the working procedures under the ESCO model, it is also noteworthy that numerous synergy effects can be generated and that the majority of transaction barriers specific to their national energy markets can be overcome. Such an integrative framework for international energy efficiency and carbon mitigation projects would contribute to the modernization of these economies' energy industries, promoting a "win-win" situation for foreign companies seeking to invest in a sustainable manner (Garbuzova and Madlener, 2012).

From the top to the bottom of the funding sponsors shown in Figure 9.2, since the positions are $x_1 - x_6$, we can replace: x_{China} = positions $[x_1, x_2, x_4]$ and x_{Europe} = positions $[x_3, x_5, x_6]$. We observed that positions x_1, x_2, x_4 are China-centered, while positions x_3, x_5, x_6, are European-centered.

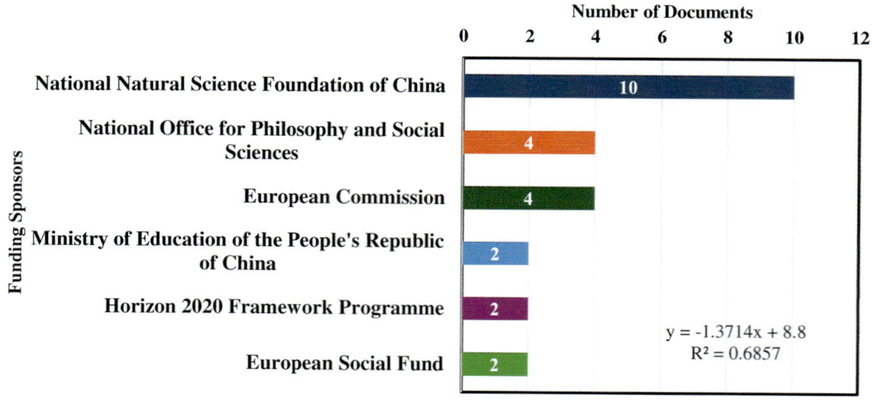

FIGURE 9.2
The top six funding sponsors in the joint fields of "low-carbon economy" and energy.

Source: Authors' own synthesis

TABLE 9.11

Percentage Allocation Among the Top Six Funding Sponsors in the Joint Fields of "Low-Carbon Economy" and Energy

Funding Sponsors	**Number of Documents in Descending Order (%)**
National Natural Science Foundation of China	41.7
National Office for Philosophy and Social Sciences	16.7
European Commission	16.7
Ministry of Education of the People's Republic of China	8.3
Horizon 2020 Framework Programme	8.3
European Social Fund	8.3
Total	**100**

Source: Authors' own synthesis

Therefore, having shaped these two fictitious geographical poles of coverage and calculating the weighted documents production in China and Europe using the linear equation that has been derived at the right side of the frame of Figure 9.2: $y = -1.3714 \times x + 8.8$, we observed that the pairs of (x,y) in China and Europe, regarding the funding sponsors allocation between them, are as follows.

$x_{China} = (1 + 2 + 4)/3 = 2.33$, and yielding the weighted number of documents in $y_{China} = 5.60$ documents

$x_{Europe} = (2 + 4 + 4)/3 = 3.33$, and yielding the weighted number of documents in $y_{Europe} = 4.23$ documents

Assuming that the funding sponsors' interest is directly analogous with the funding researches-documents, then the relevant intensity factor in these two fictitious geographical poles is: $y_{China}/y_{Europe} = 1.32$, showing that China-based funding in the research on "low-carbon economy" and energy is almost 1.32 times higher than that of Europe-based funding in the same research field. In other words, of every seven studies in the joint fields of "low-carbon economy" and energy, four are China-funded and three are European-funded.

This finding could be attributed not only to financial factors or regulations, but also to governmental motivations, initiatives, and different sizes of population, as well as to national responsibility for environmental preservation of natural sources not undermining those broadly varied socio-cultural conditions and tradition values that are nurtured and preserved among the local communities: societies and urban cities and provinces in China and Europe, respectively. In this analysis, there should be also considered two constraints: (a) the sample size of 24 documents could be considered low, considering the ongoing literature interest in the topic of low-carbon economy locally and globally; and (b) the regression coefficient is almost 0.70 – thus, a higher sample size could increase the regression coefficient value, revealing more precisely the relevant funding-publication trend between these two fictitious geographical poles. As the literature production is growing, other geographical pole(s) should emerge, making necessary the utilization of more complicated statistical tools for more close investigation, such as ANOVA, K-means, and cluster analyses.

As shown in Table 9.1,2 two-thirds of the total funding are China-sponsored while the rest one third of the total funding are Europe-sponsored. It is not a surprise that there are not other regions represented except for these two geographical areas, since it is clear that the present and the future situations of funding sponsorship in the field of low-carbon economy and energy are primarily and foremost attributed to first China, followed by Europe. The U.S. interest in carbon pollution and environmental depletion, shaping relevant motivations and actions of environmental protection, is traditionally based on which political party controls the government; thus, the governmental authorities are the major contributors of the high – or low – level of interest reported in any long-term analysis.

Based on Table 9.16, it can be denoted that low-carbon economy makes striking effect on economy and society, mainly affecting employment. It was shown that low carbon is not only a kind of life philosophy – the development

TABLE 9.12

The Top Six Subject Areas in the Joint Fields of "Low-Carbon Economy" and Environment

Subject Areas	Number of Documents in Descending Order (%)
Engineering, 7	28.1
Environmental science, 6	27.5
Energy, 5	18.1
Mathematics, 3	15.8
Social sciences, 2	10.5
Computer science, 2	10.5
Total, 25	**100**

Source: Authors' own synthesis

TABLE 9.13

The Top Six Document Types in the Joint Fields of "Low-Carbon Economy" and Environment

Document Types,#	Number of Documents in Descending Order (%)
Article, 9	47.4
Conference paper, 4	21.0
Book chapter, 2	10.5
Retracted, 2	10.5
Review, 1	5.3
Erratum, 1	5.3
Total, 19	**100**

Source: Authors' own synthesis

TABLE 9.14

The Top Three Source Titles in the Joint Fields of "Low-Carbon Economy" and Energy

Source Titles,#	Number of Documents in Descending Order (%)
Advanced Materials Research, 3	42.8
Mathematical Problems in Engineering, 2	28.6
Journal of Environmental Protection and Ecology, 2	28.6
Total, 7	**100**

Source: Authors' own synthesis

TABLE 9.15

The Top Five Clusters of Key Words in the Joint Fields of "Low-Carbon Economy" and Energy

Clusters of Key Words	Number of Documents (in Descending Order)
Carbon-Low-carbon economy, 38	Low-carbon economy, 9; low-carbon economy, 8; carbon 5, carbon dioxide 1, carbon emission, 4, carbon emission right, 1; carbon price, 1; low carbon, 2; low-carbon greens, 1; low-carbon economic, 1; low-carbon economy transition, 1; low-carbon concept, 1; low-carbon environment 1, low-carbon hospital, 1; low energy consumption, 1 (30.6%)
Environment, 27	Environmental management, 6; environmental governance, 4; ecological environment, 4; resource–environment coupling, 1; environmental accounting information disclosure, 1; environment, 1; environment management, 1; environmental analysis, 1; environmental engineering, 1; environmental-friendly interior, 1; environmental legislation, 1; environmental quality, 1; environmental regulations, 1; environment friendly, 1; enterprise environmental management, 1; national environmental legislation, 1 (21.8%)
Socio-economics, 26	Environmental economics, 5; economics, 4; economic development, 4; economic growth, 2; economic and social effects, 2; societies and institutions 1, social stability, 1; costs, 1; cost structure, 1; cost drivers, 1; cost, 1; economic strategy, 1; economic growth, 1; economic conditions, 1 (21.0%)
Regional sustainability, 22	Sustainable development, 7; agricultural waste, 1; agricultural products, 1; agricultural production, 1; agricultural pollution, 1; coordinated development, 1; coordinated growth, 1; development ability, 1; education, 1; employment, 1; employment situation, 1; employment status, 1; employment work, 1; regional analysis, 1; regional development, 1; regional resources and environment, 1 (17.7%)
Energy, 11	Renewable energy resources, 5; energy resources management, 2; electric power generation, 1; energy structures, 1; energy utilization, 1; new energy technologies, 1 (8.9%)
Total, 124	**100**

Source: Authors' own synthesis

of new energy technology rises to national economic strategy and may become a new economic growth point (Zhao and Wang, 2012). Subsequently, the adjustment of energy structure not only benefits the sustainable development road of economic development to "green," but also brings in new hope in order to improve the current employment situation. The employment situation is strongly linked to university students, thus, college graduates are

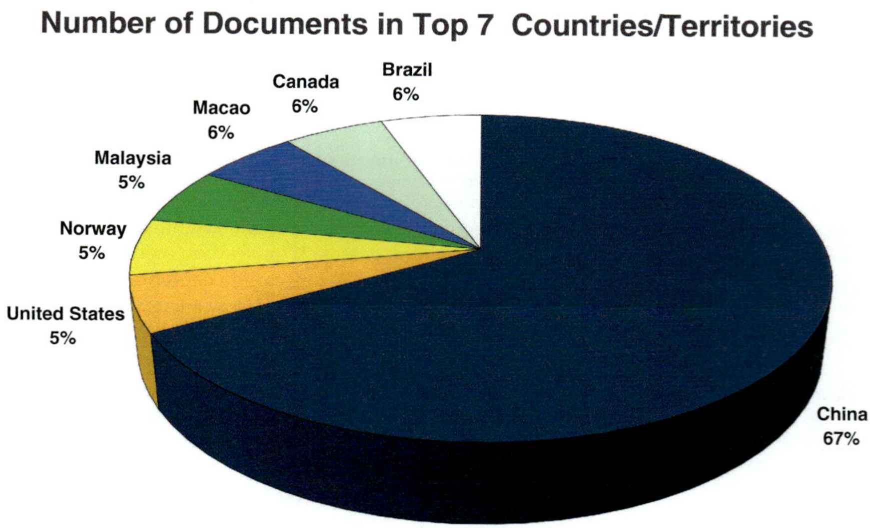

FIGURE 9.3
Percentage allocation of 12 documents reported on the top seven countries/territories in the joint fields of "low-carbon economy" and environment.

Source: Authors' own synthesis

an important index of social inspection and evaluation of the quality of running a school, being one of the core competitiveness of schools. Therefore, graduate employment status can be also directly related to the school reputation, making challenging investigation into how colleges and universities do the employment work of graduates (Zhao and Wang, 2012).

Based on Figure 9.3, it is noteworthy that the prevailing role of addressing the low-carbon energy target in the light of environmental sustainability is not new, but rather dates back to the research of almost a decade ago (Zhang and Niu, 2013). In this study, it was denoted that during its 11th Five-Year Plan, the Chinese government shifted its pollution reduction policy from simply making benchmark goals to setting binding targets with legal effects, which will push all parties involved to meet targets ahead of schedule. In practice, it has been proven that strengthening governmental responsibility and smartly appropriating tasks at various levels are two important safeguards for achieving significant progress in environmental protection during the 11th Five-Year Plan (Zhang and Niu, 2013).

Environmental protection has optimized economic development, facilitated the adjustment of economic structure and changes in development patterns, and promoted the co-development of environment, economy, and society as a whole. More importantly, a breakthrough in national environmental legislation offered a strong legal guarantee for accomplishing goals during the 11th Five-year Plan. However, we should be aware that while

TABLE 9.16

The Top Four Source/Types in the Joint Fields of "Low-Carbon Economy" and Environment

Source/Type	Number of Documents in Descending Order (%)
Journal, 12	63.2
Book series, 4	21.0
Conference proceeding, 2	10.5
Book, 1	5.3
Total, 19	**100**

Source: Authors' own synthesis

China's environmental situation has partly improved, many violations remain unchecked, and the overall outlook is still grim. Therefore, the 12th Five-year Plan further emphasizes environmental objectives. The government aims to promote environmental development using the 12th Five-Year Special Plan, which focuses on environmental issues that threaten public health and sustainable development. The plan will further promote the four strategic tasks: reducing total emissions, strengthening the improvement of environmental quality, preventing environmental risks, and perfecting basic environmental public services (Zhang and Niu, 2013).

9.4 Discussion

Based on the yielded documents, it is noteworthy that the rapid growth of social and economic development has also brought great challenges to the development of regional resources and the environment. The deepening and broadening of the sustainable development concept implies the coordinated development of regional resources, the environment, and economic growth. Indeed, the coordinated development of resources, the environment, and economic growth is not considered only as the core of industrial growth, but also the promotional accelerator of industrial transfer and the optimal allocation of resources are unavoidably fostering and nurturing factors of social development, as seen in Table 9.17.

It cannot be overstated that the necessity of steady upgrading of the industrial structure has subtly revealed severe problems, such as environmental pollution, unbalanced economic development, and resource shortages, all affecting on various important development areas. The reported findings of this study are consistent with the broadly agreed statement that the development of a low-carbon economy has strong feasibility to promote the

TABLE 9.17

The Open Access Characteristics of the Yielded Documents in the Joint Fields of "Low-Carbon Economy" and Energy and Environment

Open Access "Low-Carbon Economy" and Energy, #	Open Access "Low-Carbon Economy" and Environment, #	Number of Documents (in Descending Order and in: ["Environment"/"Energy"] × 100%
Gold, 22	Gold, 4	18.2%
Green, 14	No data reported	–
Bronze, 5	No data reported	–
Hybrid gold, 3	Hybrid gold, 1	33.3%
All open access, 35	All open access, 5	14.3%

Source: Authors' own study

coordinated development of regional resources, the environment, and economic growth.

Based on the aforementioned studies, it can be argued that the amount of energy directly determines the speed and sustainability of economic development, but the consumption of energy is unavoidably a polluter of the environment. The study of Chi et al. (2019) included five major scenarios: benchmarks, different economic growth rates, different industrial structures, energy conservation, and integration. The study underscored that the energy consumption of the tertiary industry can continue to increase, and the growth rate is apparent (Chi et al., 2019).

The energy resources of the economies play a critical role in reaching the 2030 and 2050 decarbonization targets. In this respect, research and development (R&D) budgets for nuclear energy (NRD), which constitutes one-fifth of the current energy resources in the UK, can be a step in the transition to a low-carbon economy. In this context, while positive changes in NRD reduce environmental degradation, negative changes in NRD hinder environmental sustainability (Caglar, 2023). It was also notable that while economic growth and trade openness reduce environmental quality, financial development contributes to environmental quality (Caglar, 2023). Considering this, deepening financial activities should continue at least at national scales of investigation, while more resources should be allocated to nuclear energy R&D budgets in the transition to a low-carbon economy, especially among Western developed and highly industrialized economies (Caglar, 2023).

A technological-based constraint is that the intermittency and uncertainty of new energy – especially that which has been yielded from renewable energy sources (RES) – can easily lead to a mismatch between the supply and demand of energy, which limits the wide application of integrated energy systems (IES). Based on these considerations, an optimal scheduling model for an IES was proposed in the literature, considering the bilateral interaction mechanism between supply and demand and the reward-penalty ladder type of carbon trading (Yi and Ren, 2024). The introduction of a reward-penalty ladder type of carbon trading mechanism was decided, while the optimal dispatching model can be established to minimize the total cost of the IES (Yi and Ren, 2024). The dual response of supply and demand can not only increase the coordination and optimization ability of the IES but also achieve the improvement of the system economy and promote the consumption of renewable energy based on ensuring the comfort of users (Yi and Ren, 2024). The reward-penalty ladder type of carbon trading model can restrain the carbon emissions of the system to a greater extent and synergize the low carbon and economy of the system (Yi and Ren, 2024).

According to the IES studies, it was also argued that most of the existing studies of the low-carbon operation of integrated energy systems focus on the source side and rarely extend to the load side, and do not consider the DR characteristics of different loads (Liu et al., 2025). Therefore, based on the carbon flow tracing method of the power system, a model was applied to adjust the load side operating state of the power system by using price incentive (Liu et al., 2025). This research approach has been motivated by the gradual liberalization of the carbon and distributed trading markets; thus, the economic incentive trading market mechanism has become an effective way to promote carbon emission reduction in microgrids (Liu et al., 2025). The simulation results showed that the model combines carbon trading and DR, which can effectively reduce carbon emissions and significantly improve the environmental benefits of the system (Liu et al., 2025).

From a financial perspective, the examination of the effect of stock market development (SMD) on the low-carbon economy (LCE) is noteworthy (Zeqiraj et al, 2020). The authors consider two channels by which this affect occurs: renewable energy and technological innovation. Focusing on EU member nations, it was demonstrated that SMD impedes LCE in the long run. In contrast, technological innovation (TI) was found to be a driving factor in achieving LCE in the long run (Zeqiraj et al, 2020). These results also supported the argument that renewable energy consumption and production enhance LCE. Stock market development fosters LCE through the channels of renewable energy and technological innovation (Zeqiraj et al, 2020). In another financial-based study (Habiba et al., 2023), the authors stressed that global warming has become one of the most serious and hotly debated global issues in recent years, despite increasing international agreements requiring urgent action. Therefore, many countries are striving to achieve carbon neutrality. To achieve decarbonization, renewable energy and green technologies can play a significant role. Thus, it is an imperative need to investigate the connections among renewable energy consumption, green technology, and carbon emissions (Habiba et al., 2023). In response to this emerging issue, the authors investigated the effects of financial development (FD), renewable energy consumption, and green technology on carbon emissions (CE) in seven emerging countries between 1990 and 2020 (Habiba et al., 2023). To assess the empirical data, a wide range of econometric techniques were employed, and these empirical findings revealed that FD increased the level of CE, which resulted in environmental degradation. In contrast, renewable energy use and green technology declined CE in the long run. Furthermore, when combined with renewable energy, FD tended to be less detrimental to the environment. FD improved environmental quality through the green technology channel. The findings of causality tests demonstrated that policies related to renewable energy and green technology can affect carbon emissions in one direction, but not in the other (Habiba et al., 2023). In this context, it is crucial to promote green technology and renewable energy use, especially in emerging economies, enabling attainment of sustainable development goals (Habiba et al., 2023).

From a geographical perspective, as countries scramble for cleaner energy production and to meet carbon reduction targets, natural gas was proposed by Mozakka et al. (2022) to become an increasingly attractive option, with liquefied natural gas (LNG) as a popular transportation choice. Besides determining those various factors affecting the global natural gas market, its recent history, current state, and future, having a closer look at the possibility of East Asia becoming an alternative market to Europe for Russian LNG, authors bring in the United States both as a political force that employs economic sanctions and as a potential LNG supplier. Defining a three-player game involving Russia, the United States, and Japan resulted in relative market stability. Moreover, in the case of sanctions against Russia, authors mentioned that it will lose its foothold in the Japanese market in the long term. Additionally, the study is an attempt of the potential of LNG trading as the first step for East Asia's energy transition to a low-carbon economy (Mozakka et al., 2022).

Based on the literature search and the yielded documents in this study, it can be also denoted the importance of conducting studies that integrate: (a) electricity infrastructure, which is a driver of economic and technological urban development, especially in emerging economies for the electricity mix in urban areas of the country (de Magalhães et al., 2023); and (b) promotion of low-carbon technologies, such as renewable energy or carbon capture and storage (CCS), in which the benefits of adapting to a low-carbon economy can be assessed and technological adaptation strategies can be valued in alignment with associated impacts on human health and the environment (Thorne, 2016). Such an integrated research approach can be undertaken under a life cycle assessment (LCA). Indeed, CCS that has focused on LCA was proven a particularly useful tool to support a holistic view of the impacts relevant to mitigation options. These are important decision criteria for policymakers to be dealt with (Thorne, 2016). Regarding the electricity production, free energy consumers' strategic role as agents in emerging countries' public policies can promote renewable sources in regional and national contexts. In these contexts, life cycle characteristics of urban electrical solutions are beneficial among emerging countries, and they influence regional and national sustainable energy management. Therefore, the LCA studies should provide the integration of sustainable and technical elements of electricity/CCS systems turning the managerial decision-making towards the energy transition more robust (de Magalhães et al., 2023; Liu and Yang, 2013).

9.5 Conclusions

Based on this study, some priority actions and designing policies that could systematize the "low-carbon economy" research in a more stable and prosperous routes of development have emerged. In this context, the actions and policies of future research interest can be proposed.

- Empirical studies can be carried out to evaluate and analyze the relationships among green financial behaviour, climate change mitigation, and environmental energy sustainability. Besides these analyses, estimates can be made regarding the sustainability of the natural world and economy. Data collected from large-scaling surveys from respondents can be tested against the research models, while descriptive statistics of compiled data from the completed surveys can be calculated when appropriate for liable and credible outcomes to be derived (Ionescu, 2021).
- A low-carbon future will be good for renewable energy under both the development pathways, though the share of renewable energy will be more and more higher under a sustainable pathway. Renewable energy faces competition from low-carbon technologies like nuclear and carbon capture and storage in the electricity sector. Solar, wind, biomass, and biofuels are fast emerging as the four competitive renewable energy choices for densely populated places of the planet such as India. Renewable development, however, depends critically on reduction in the costs and in the ability to integrate the intermittent renewables within the existing systems for which technology transfer and capacity building hold the key (Shukla et al., 2010).
- From a geographical and spatial perspective, it is noteworthy that with the acceleration of urbanization, cities are the main targets for carbon neutrality and urban energy is the terminal of energy consumption and the integration point of various energy systems. Therefore, there is a need to promote the development of urban green energy and achieve low input and high output to achieve a low-carbon economy in cities (Wu et al., 2022). While earlier published studies neglected to consider the input–output efficiency of

urban green energy development, present and future studies can be based on improved data-envelopment analysis, making a comparative evaluation of the input–output efficiency of green energy development, especially in densely populated urban areas (Wu et al., 2022). Considering the differences in regions, the development of urban green energy in different provinces can be classified. From the perspective of a low-carbon economy, economic growth factors and environmental constraint factors are best considered as case-specific setting variables (Wu et al., 2022). The research objectives of such types of study is the identification of the input–output efficiencies of urban green energy development of local provinces or cities of relatively high levels. Provinces with faster economic development and higher environmental carrying capacity can show advantages after optimization and to become pilot areas for the development of urban green energy (Wu et al., 2022). This research should model the development of urban green energy in various provinces from the input and output perspective (Wu et al., 2022).

- Considering the fact that urban development faces resource shortages, environmental pollution, and ecological crises, the development of a low-carbon economy can promote the sustainable development of cities. On the other hand, the development of low-carbon cities based on sustainable development can improve and upgrade the successful development of low-carbon cities (Wang et al., 2021). In studies such as these, the comprehensive evaluation system for low-carbon cities can be established from a sustainable development perspective. The evaluation system for low-carbon cities can characterize the following four sub-systems of "development": economic, social, low-carbon, and sustainable (Wang et al., 2021). Using the entropy weight method then enables the determination of the weight of each subsystem and the calculation of the low-carbon comprehensive development level score of regions in a time span of one or more decades. These calculation results disclosed a low-carbon comprehensive development level to be determined by the region and the timespan examined (Wang et al., 2021).
- From a sectoral perspective, the significance of the low-carbon economy in the transportation sector can be seen whereby low-carbon green cars terminal return logistics operation modes can compare and analyze various low-carbon types of green cars in order to better handle situations and to resolve problems that put forward better solutions for a short-term or local level of analysis (Lv, 2013).

Conclusively, it can be underscored that a low-carbon economy is one concerned with long-term stability and one which demonstrates an awareness of the interactions between the economy, society, and the environment. The science of climate change is clear enough to signal that the export opportunities for any economy that moves in this direction will be huge (Hewett, 2002). Many of the solutions involve rapid technological change, innovation, and improved productivity. However, innovation or economic changes are related to vested interests bent on preventing change and delaying progress in order to protect short-term interests. Some resources will have to be devoted to helping these economic actors adapt to the new economy. Under these conditions, failing to move towards a low-carbon economy now will make it much tougher – both politically and economically – to shift to one in the future (Hewett, 2002).

References

Ali E.B., Gyamfi B.A., Kwakwa P.A., Agbozo E. (2024). Transitioning to low carbon economy among OECD countries: Do renewable energy, globalization and higher economic growth matter? *Energy and Environment*, 35 (8), 4248–4271. https://doi.org/10.1177/0958305X231177746

Alonso P.M., Hewitt R., Pacheco J.D., Bermejo L.R., Jiménez V.H., Guillén J.V., Bressers H., de Boer C. (2016). Losing the roadmap: Renewable energy paralysis in Spain and its implications for the EU low carbon economy. *Renewable Energy*, 89, 680–694. https://doi.org/10.1016/j.renene.2015.12.004

Alsiyabi M., Hassan H., Khudari M. (2024). Evaluating the role of GCC nations in sustainable energy investment using an enhanced low-carbon economy index. *Journal of Ecohumanism*, 3 (7), 3205–3215. https://doi.org/10.62754/joe.v3i7.4462

Bridge G., Bouzarovski S., Bradshaw M., Eyre N. (2013). Geographies of energy transition: Space, place and the low-carbon economy. *Energy Policy*, 53, 331–340. https://doi.org/10.1016/j.enpol.2012.10.066

Caglar A.E. (2023). Can nuclear energy technology budgets pave the way for a transition toward low-carbon economy: Insights from the United Kingdom. *Sustainable Development*, 31 (1), 198–210. https://doi.org/10.1002/sd.2383.

Chang M.-C. (2015). Room for improvement in low carbon economies of G7 and BRICS countries based on the analysis of energy efficiency and environmental Kuznets curves. *Journal of Cleaner Production*, 99, 140–151. https://doi.org/10.1016/j.jclepro.2015.03.002

Chen C., Wu Q., Xin H. (2019). Carbon balance of marine energy based on carbon emission and low carbon economy: A case study of Shandong province. *Journal of Coastal Research*, 98, 167–170. https://doi.org/10.2112/SI98-041.1

Chen J., Li H. (2024). Impact of energy storage industry development on the low-carbon economy: Evidence from China. *Frontiers in Energy Research*, 12, 1413057. https://doi.org/10.3389/fenrg.2024.1413057

Chi Y., Yuan L., Li H., Zhang Y., Bai G. (2019). Using LEAP model to predict energy consumption of Beijing under the constraint of low-carbon economy. *Ekoloji*, 28 (107), 107141, 1205–1211.

de Magalhães R.F., Danilevicz Â.D.M.F., Passuello A., Echeveste M.E. (2023). Exploring life cycle trade-offs of the electricity solutions on Brazilian urban environments: Ways towards a low-carbon economy. *Sustainable Production and Consumption*, 36, 425–438. https://doi.org/10.1016/j.spc.2023.01.019

Dembicka-Niemiec A., Szafranek-Stefaniuk E., Kalinichenko A. (2023). Structural and investment funds of the European Union as an instrument for creating a low-carbon economy by selected companies of the energy sector in Poland. *Energies*, 16 (4), 2031. https://doi.org/10.3390/en16042031

Ding C., Liu M.-Y., Wan S.-Z., Cai Z.-Y., Yuan L.-J., Wang R., Zheng P. (2023). The sustainable treatment of chemical polishing wastewater under low-carbon economy: Pollutants removal, resource recovery and energy generation. *Journal of Cleaner Production*, 414, 137469. https://doi.org/10.1016/j.jclepro.2023.137469

Field R.A., Derwent R.G. (2021). Global warming consequences of replacing natural gas with hydrogen in the domestic energy sectors of future low-carbon economies in the United Kingdom and the United States of America. *International Journal of Hydrogen Energy*, 46 (58), 30190–30203. https://doi.org/10.1016/j.ijhydene.2021.06.120

Gao D. (2025). A deep learning approach for energy management systems in smart buildings towards a low-carbon economy. *International Journal of Low-Carbon Technologies*, 20, 1136–1142. https://doi.org/10.1093/ijlct/ctaf063

Garbuzova M., Madlener R. (2012). Towards an efficient and low carbon economy post-2012: Opportunities and barriers for foreign companies in the Russian energy market. *Mitigation and Adaptation Strategies for Global Change*, 17 (4), 387–413. https://doi.org/10.1007/s11027-011-9332-8

Giber V.O. (2019). The energy policy of the European Union and China toward the Arctic in view of falling oil and gas prices, climate change and low-carbon economies. *Ecocycles*, 5 (2), 80–96. https://doi.org/10.19040/ecocycles.v5i2.155

Gulzar M.A., Asghar H., Hwang J., Hassan W. (2020). China's pathway towards solar energy utilization: Transition to a low-carbon economy. *International Journal of Environmental Research and Public Health*, 17 (12), 4221, 1–11. https://doi.org/10.3390/ijerph17124221

Gunningham N., Bowman M. (2016). Energy regulation for a low carbon economy: Obstacles and opportunities. *Environmental and Planning Law Journal*, 33 (2), 118–141.

Habiba U., Xinbang C., Ali S. (2023). Investigating the impact of financial development on carbon emissions: Does the use of renewable energy and green technology really contribute to achieving low-carbon economies? *Gondwana Research*, 121, 472–485. https://doi.org/10.1016/j.gr.2023.05.013.

He K., Zeng L., Yang J., Gong Y., Zhang Z., Chen K. (2024). Optimization strategy for low-carbon economy of integrated energy system considering carbon capture-two stage power-to-gas hydrogen coupling. *Energies*, 17 (13), art. no. 3205. https://doi.org/10.3390/en17133205

Hewett C. (2002). Priorities for a low-carbon economy: The spending review and the environment. *New Economy*, 9 (1), 11–15. https://doi.org/10.1111/1468-0041.00232

Hosein A.S., Whale J., Simsek Y., Urmee T. (2023). Exploring energy policy scenarios to transition to a low carbon economy by 2050: A case study on the Northern Territory of Australia. *Energy Policy*, 180, art. no. 113663, Cited 17 times. https://doi.org/10.1016/j.enpol.2023.113663

Hou J., Zhang P., Tian Y., Yuan X., Yang Y. (2011). Developing low-carbon economy: Actions, challenges and solutions for energy savings in China. *Renewable Energy*, 36 (11), 3037–3042. https://doi.org/10.1016/j.renene.2011.03.033

Hou W., Pan X., Pu Y., Ma R., Saleem A. (2024). Harnessing artificial intelligence for energy strategies: Advancing global economic policies and hydrogen production in the transition to a low-carbon economy. *Energy Strategy Reviews*, 56, 101568. https://doi.org/10.1016/j.esr.2024.101568

Huang X., Yi H., Liu M. (2024). Development of a low-carbon economy with the dual transformation of economy and energy structure. *International Journal of Global Energy Issues*, 46 (3–4), 365–388. https://doi.org/10.1504/IJGEI.2024.137091

Ionescu L. (2021). Transitioning to a low-carbon economy: Green financial behavior, climate change mitigation, and environmental energy sustainability. *Geopolitics, History, and International Relations*, 13 (1), 86–96. https://doi.org/10.22381/GHIR13120218.

Ishida H. (2018). Can nuclear energy contribute to the transition toward a low-carbon economy? The Japanese case. *International Journal of Energy Economics and Policy*, 8 (1), 62–68.

Jiang B., Sun Z., Liu M. (2010). China's energy development strategy under the low-carbon economy. *Energy*, 35 (11), 4257–4264. https://doi.org/10.1016/j.energy.2009.12.040

Kanchana K. (2024). Comparative approaches to energy transition: Policy guideline for enhancing Thailand's path to a low-carbon economy. *Energies*, 17 (22), 5620. https://doi.org/10.3390/en17225620

Kapitonov I.A. (2020). Low-carbon economy as the main factor of sustainable development of energy security. *Industrial Engineering and Management Systems*, 19 (1), 3–13. https://doi.org/10.7232/iems.2020.19.1.003

Kapitonov I.A. (2021). Development of low-carbon economy as the base of sustainable improvement of energy security. *Environment, Development and Sustainability*, 23 (3), 3077–3096. https://doi.org/10.1007/s10668-020-00706-0

Kitasei S. (2010). Powering the low-carbon economy: The once and future roles of renewable energy and natural gas. *Worldwatch Paper*, (184), 1–46.

Knuth S. (2023). Rentiers of the low-carbon economy? Renewable energy's extractive fiscal geographies. *Environment and Planning A*, 55 (6), 1548–1564. https://doi.org/10.1177/0308518X211062601

Krishnan V., McCalley J.D. (2016). The role of bio-renewables in national energy and transportation systems portfolio planning for low carbon economy. *Renewable Energy*, 91, 207–223. https://doi.org/10.1016/j.renene.2016.01.052

Kuznetsov V.V., Gasenko O.A. (2023). Thermophysical and physicochemical basis for hydrogen production in compact energy systems of low-carbon economy. *Thermophysics and Aeromechanics*, 30 (5), 949–954. https://doi.org/10.1134/S086986432305013X

Lechtenböhmer S., Schneider C., Roche M.Y., Höller S. (2015). Re-industrialisation and low-carbon economy-can they go together? Results from stakeholder-based scenarios for energy-intensive industries in the German state of North Rhine Westphalia. *Energies*, 8 (10), 11404–11429. https://doi.org/10.3390/en81011404

Li A., Haron S.H.B. (2024). Evaluation of sustainable environment-friendly interior decoration design from the perspective of low-carbon economy. *Journal of the Balkan Tribological Association*, 30 (2), 260–274.

Li J., Khan I. (2024). Transition to a sustainable energy balance and conversion factors in economic management: Basics for transitioning to a low-carbon economy. *Energy*, 313, 133724. https://doi.org/10.1016/j.energy.2024.133724

Li T., Shi Z., Han D. (2021). Does renewable energy consumption contribute to the development of low-carbon economy? Evidence from China. *Environmental Science and Pollution Research*, 28 (39), 54891–54908. https://doi.org/10.1007/s11356-021-14468-3

Li X., Gao H., Zhou E. (2024). Research on optimization of financial performance evaluation of energy enterprises under the background of low-carbon economy. *Energies*, 17 (10), 2311. https://doi.org/10.3390/en17102311

Li X., Gao J. (2023). Decision-making model of sustainable ecological environment management based on enterprise environmental management under a low-carbon economy. *Frontiers in Energy Research*, 11, 1221910. https://doi.org/10.3389/fenrg.2023.1221910

Li Y., Dai G. (2019). Energy saving and emission reduction of fossil energy based on low carbon economy and its consumption structure optimization. *International Journal of Low-Carbon Technologies*, 14 (3), 381–385. https://doi.org/10.1093/ijlct/ctz025

Liang X., Xu J. (2024). Accelerating transition to a low-carbon economy: A coupling analysis of agricultural products and resource environment. *Sustainability*, 16 (15), 6315. https://doi.org/10.3390/su16156315

Lingjun L., Chao W. (2021). Management of hospital low carbon operation environment under the concept of low carbon economy. *Journal of Environmental Protection and Ecology*, 22 (1), 433–442.

Liu C., Wang W., Li J., Liu X., Chi Y. (2024). Optimized scheduling of integrated energy systems for low carbon economy considering carbon transaction costs. *Global Energy Interconnection*, 7 (4), 377–390. https://doi.org/10.1016/j.gloei.2024.08.010

Liu L., Yao X., Qi X., Han Y. (2023). Low-carbon economy configuration strategy of electro-thermal hybrid shared energy storage in multiple multi-energy microgrids considering power to gas and carbon capture system. *Journal of Cleaner Production*, 428, 139366. https://doi.org/10.1016/j.jclepro.2023.139366

Liu Y., Wang X., Li S., Liu L., Zhao Y., Yu M. (2025). Demand response model of low-carbon economy in integrated energy system based on carbon flow traceability. *Engineering Reports*, 7 (4), 70097. https://doi.org/10.1002/eng2.70097.

Liu Z., Yang R. (2013). Using the particle swarm optimization model to evaluate the wind power enterprise development ability under low-carbon economy environment. *Advanced Materials Research*, 608–609, 683–686. https://doi.org/10.4028/www.scientific.net/AMR.608-609.683

Lv Y.F. (2013). Based on low carbon economy under the environment of green car terminal returned logistics is analysed. *Advanced Materials Research*, 779, 1543–1546. https://doi.org/10.4028/www.scientific.net/AMR.779-780.1543

Martí-Ballester C.-P. (2019). Do European renewable energy mutual funds foster the transition to a low-carbon economy? *Renewable Energy*, 143, 1299–1309. https://doi.org/10.1016/j.renene.2019.05.095

Martínez-Martínez S., Pérez-Villarejo L., Eliche-Quesada D., Sánchez-Soto P.J. (2023). New Types and dosages for the manufacture of low-energy cements from raw materials and industrial waste under the principles of the circular economy and low-carbon economy. *Materials*, 16 (2), 802. https://doi.org/10.3390/ma16020802

Mozakka M., Salimi M., Hosseinpour M., N. Borhani T. (2022). Why LNG can be a first step in East Asia's energy transition to a low carbon economy: Evaluation of challenges using game theory. *Energies*, 15 (17), 6476. https://doi.org/10.3390/en15176476.

Murshed M., Ahmed Z., Alam M.S., Mahmood H., Rehman A., Dagar V. (2021). Reinvigorating the role of clean energy transition for achieving a low-carbon economy: Evidence from Bangladesh. *Environmental Science and Pollution Research*, 28 (47), 67689–67710. https://doi.org/10.1007/s11356-021-15352-w

Nguyen, V. G., Sirohi, R., Tran, M. H., Truong, T. H., Duong, M. T., Pham, M. T., Cao, D. N. (2024). Renewable energy role in low-carbon economy and net-zero goal: Perspectives and prospects. *Energy and Environment*. https://doi.org/10.1177/0958305X241253772

Nhamo G., Ho S.-Y. (2011). Renewable energy policy landscape in South Africa: Moving towards a low carbon economy. *WIT Transactions on Ecology and the Environment*, 143, 265–276. https://doi.org/10.2495/ESUS110231

Ni W., Chen Z. (2011). Synergistic utilization of coal and other energy – Key to low carbon economy. *Frontiers of Energy and Power Engineering in China*, 5 (1), 1–19. https://doi.org/10.1007/s11708-010-0136-4

Pan Y., Dong F. (2023). Green finance policy coupling effect of fossil energy use rights trading and renewable energy certificates trading on low carbon economy: Taking China as an example. *Economic Analysis and Policy*, 77, 658–679. https://doi.org/10.1016/j.eap.2022.12.014

Porshnov D. (2022). Evolution of pyrolysis and gasification as waste to energy tools for low carbon economy. *Wiley Interdisciplinary Reviews: Energy and Environment*, 11 (1), 421. https://doi.org/10.1002/wene.421

Robinson L. (2010). Linking transformational materials and processing for an energy-efficient and low-carbon economy. *JOM*, 62 (3), 11–15. https://doi.org/10.1007/s11837-010-0040-8

Romano A.A., Scandurra G. (2011). The investments in renewable energy sources: Do low carbon economies better invest in Green technologies? *International Journal of Energy Economics and Policy*, 1 (4), 107–115.

Sagastume Gutiérrez A., Cabello Eras J.J., Huisingh D., Vandecasteele C., Hens L. (2018). The current potential of low-carbon economy and biomass-based electricity in Cuba. The case of sugarcane, energy cane and marabu (*Dichrostachys cinerea*) as biomass sources. *Journal of Cleaner Production*, 172, 2108–2122. https://doi.org/10.1016/j.jclepro.2017.11.209

Serrano-Puente D. (2021). Are we moving toward an energy-efficient low-carbon economy? An input–output LMDI decomposition of CO2 emissions for Spain and the EU28. *SERIEs*, 12 (2), 151–229. https://doi.org/10.1007/s13209-020-00227-z

Shen L., Sun Y. (2016). Review on carbon emissions, energy consumption and low-carbon economy in China from a perspective of global climate change. *Journal of Geographical Sciences*, 26 (7), 855–870. https://doi.org/10.1007/s11442-016-1302-3

Shukla P.R., Dhar S., Fujino J. (2010). Renewable energy and low carbon economy transition in India. *Journal of Renewable and Sustainable Energy*, 2 (3), 031005. https://doi.org/10.1063/1.3411001

Shi C., Feng X., Jin Z. (2022). Sustainable development of China's smart energy industry based on artificial intelligence and low-carbon economy. *Energy Science and Engineering*, 10 (1), 243–252. https://doi.org/10.1002/ese3.856

Słupik S., Kos-łabędowicz J., Trzęsiok J. (2021). Energy-related behaviour of consumers from the Silesia province (Poland) – towards a low-carbon economy. *Energies*, 14 (8), 2218. https://doi.org/10.3390/en14082218

Soto G.H., Nghiem X.-H., Martinez-Cobas X. (2025). Analyzing the role of main energy transition policies upon renewable energy penetration in the EU: An assessment of energy productivity and low carbon economies. *Environmental and Sustainability Indicators*, 25, 100573. https://doi.org/10.1016/j.indic.2024.100573

Sun C., Miao B. (2022). Performance evaluation of reducing consumption of energy in the Yangtze River Delta under the background of low-carbon economy. *Journal of Mathematics*, 3235776. https://doi.org/10.1155/2022/3235776

Sun X., Onuh W.O. (2024). Regional energy consumption policy practice to promote low-carbon economy. *Frontiers in Environmental Science*, 12, 1423009. https://doi.org/10.3389/fenvs.2024.1423009

Tariq A., Nawaz F., Haider S.A., El Rafae G.A., Kayani U., Aziz A.L. (2025). Transitioning towards low carbon economy: Role of renewable energy, economic growth and FDI in achieving SDGs. *International Journal of Energy Economics and Policy*, 15 (3), 403–410. https://doi.org/10.32479/ijeep.18476

Thorne R.J. (2016). Transition to a low carbon economy; impacts to health and the environment. *Molecular and Integrative Toxicology*, 169–201. https://doi.org/10.1007/978-3-319-43142-0_9

Ullah K., Abbas S., Tariq M., Mahmood N., Kaechele H. (2023). The symmetric and asymmetric impacts of green energy, eco-innovation, and urbanization in explaining low-carbon economy for Pakistan. *Environmental Science and Pollution Research*, 30 (12), 33375–33395. https://doi.org/10.1007/s11356-022-24407-5

Wang T., Du Y., Yao D. (2021). Comprehensive evaluation of sustainable development of low-carbon economy and environment in Anhui province. *Fresenius Environmental Bulletin*, 30 (2 A), 2147–2158.

Wang X. (2022). Evaluation of environmental accounting information disclosure level of manufacturing industry in a low-carbon economy environment. *Journal of Environmental Protection and Ecology*, 23 (1), 389–400.

Wang Y., Huang J. (2022). Pathway to develop a low-carbon economy through energy-substitution technology in China. *Energy*, 261, 124964. https://doi.org/10.1016/j.energy.2022.124964

Wang Z. (2023). Coordinated development of regional resources, environment, and economic growth under the background of low-carbon economy. *Frontiers in Ecology and Evolution*, 11, 1181990. https://doi.org/10.3389/fevo.2023.1181990

Watanabe S., Nishimura M. (2013). Japan's experience of energy saving and business model for low-carbon economy. Chinese *Journal of Population Resources and Environment*, 11 (4), 288–298.https://doi.org/10.1080/10042857.2013.800674

Weber G., Cabras I. (2017). The transition of Germany's energy production, green economy, low-carbon economy, socio-environmental conflicts, and equitable society. *Journal of Cleaner Production*, 167, 1222–1231. https://doi.org/10.1016/j.jclepro.2017.07.223

Winkler H., Marquand A. (2009). Changing development paths: From an energy-intensive to low-carbon economy in South Africa. *Climate and Development*, 1 (1), 47–65. https://doi.org/10.3763/cdev.2009.0003

Wu C. (2025). Carbon emission evaluation and low carbon economy optimization scheduling of rural integrated energy system based on LCA method. i, 13, 17182–17194. https://doi.org/10.1109/ACCESS.2025.3533099

Wu H., Wang L., Peng D., Liu B. (2022). Input-output efficiency model of urban green-energy development from the perspective of a low-carbon economy. *Clean Energy*, 6 (1), 905–916. https://doi.org/10.1093/ce/zkab061.

Xie X., Zhao D., Cai Z. (2012). New problem research of enterprise operation in low carbon economy environment. *Advances in Information Sciences and Service Sciences*, 4 (15), 289–295. https://doi.org/10.4156/AISS.vol4.issue15.35

Xiong C., Yang D., Huo J., Zhao Y. (2015). The relationship between energy consumption and economic growth and the development strategy of a low-carbon economy in Kazakhstan. *Journal of Arid Land*, 7 (5), 15, 706–715. https://doi.org/10.1007/s40333-015-0132-y

Yang X., Niu D. (2018). Planning for incorporation of new energy sources into the electric power grid to develop the low carbon economy. *IPPTA: Quarterly Journal of Indian Pulp and Paper Technical Association*, 30 (1), 421–428.

Yi T., Ren W. (2024). Low carbon economy scheduling of integrated energy system considering the mutual response of supply and demand. *Sustainable Energy, Grids and Networks*, 38, 101279. https://doi.org/10.1016/j.segan.2024.101279.

Yu W., Huafeng W. (2019). Neural network model for energy low carbon economy and financial risk based on PSO intelligent algorithms. *Journal of Intelligent and Fuzzy Systems*, 37 (5), 6151–6163. https://doi.org/10.3233/IFS-179199

Zeqiraj V., Sohag K., Soytas U. (2020). Stock market development and low-carbon economy: The role of innovation and renewable energy. *Energy Economics*, 91, 104908. https://doi.org/10.1016/j.eneco.2020.104908.

Zhang C., Kuang Y. (2024). Low-carbon economy optimization of integrated energy system considering electric vehicles charging mode and multi-energy coupling. *IEEE Transactions on Power Systems*, 39 (2), 3649–3660. https://doi.org/10.1109/TPWRS.2023.3280067

Zhang P., Niu H. (2013). Environment governance and low carbon economy in China. *Green Low-Carbon Development in China*, 9783319011530, 127–149. https://doi.org/10.1007/978-3-319-01153-0_6

Zhang S., Wang J. (2025). The correlation between low-carbon economy and energy technology innovation in the Yangtze River Delta region under spatial econometric methods. *Frontiers in Environmental Science*, 13, 1543124. https://doi.org/10.3389/fenvs.2025.1543124

Zhang Y., Wang W., Liang L., Wang D., Cui X., Wei W. (2020). Spatial-temporal pattern evolution and driving factors of China's energy efficiency under low-carbon economy. *Science of the Total Environment*, 739, 140197. https://doi.org/10.1016/j.scitotenv.2020.140197

Zhao H., Wang H. (2012). How to do well graduate employment of colleges and universities under the low carbon economy environment. *Advanced Materials Research*, 573–574, 821–825. https://doi.org/10.4028/www.scientific.net/AMR.573-574.821

Zhao S., He G., Zheng Z., Zhang S., Wu T., Wang S., Liu X. (2024). Research on the coupling of regional low-carbon economy and energy structure based on multi algorithm optimization: A case study of Anhui Province. *Polish Journal of Environmental Studies*, 33 (4), 4455–4465. https://doi.org/10.15244/pjoes/182881

Zheng S., Huang G., Luo Z. (2025). Coordinated control strategies for polymorphic energy in hydrogen-integrated virtual power plants under the goal of a low-carbon economy. *Energies*, 18 (6), 1351. https://doi.org/10.3390/en18061351

Zhou L. (2014). Energy efficiency and investments in low-carbon economy: The impact of carbon finance on sustainability development. *Journal of Chemical and Pharmaceutical Research*, 6 (5),1255–1261.

Zhu X., Xue J., Hu M., Liu Z., Gao X., Huang W. (2023). Low-carbon economy dispatching of integrated energy system with P2G-HGT coupling wind power absorption based on stepped carbon emission trading. *Energy Reports*, 10, 1753–1764. https://doi.org/10.1016/j.egyr.2023.08.023

Zlatinov D., Nedev B., Atanasov I., Kosev N. (2019). Effects on the economic growth in Bulgaria during the transition to low-carbon economy in the energy sector. *Ikonomicheski Izsledvania*, 28 (6), 110–127.

Index

Note: Page numbers in *italics* indicate a figure and page numbers in **bold** indicate a table on the corresponding page.

W

X

For Product Safety Concerns and Information please contact our EU representative GPSR@taylorandfrancis.com Taylor & Francis Verlag GmbH, Kaufingerstraße 24, 80331 München, Germany